THE
SOCIAL
LIVES OF
ANIMALS

THE
SOCIAL
LIVES OF
ANIMALS

ASHLEY WARD

BASIC BOOKS

New York

Basic Books
Hachette Book Group
1290 Avenue of the Americas, New York, NY 10104
www.basicbooks.com

Printed in the United States of America

First U.S. Edition: March 2022

Published by Basic Books, an imprint of Perseus Books, LLC, a subsidiary of Hachette Book Group, Inc. The Basic Books name and logo is a trademark of the Hachette Book Group.

The Hachette Speakers Bureau provides a wide range of authors for speaking events. To find out more, go to www.hachettespeakersbureau.com or call (866) 376-6591.

The publisher is not responsible for websites (or their content) that are not owned by the publisher.

Print book interior design by Amy Quinn

Library of Congress Cataloging-in-Publication Data
Names: Ward, Ashley, author.
Title: The social lives of animals / Ashley Ward.
Description: First U.S. edition. | New York : Basic Books, 2022. | Includes bibliographical references and index.
Identifiers: LCCN 2021024308 | ISBN 9781541600836 (hardcover) | ISBN 9781541600843 (ebook)
Subjects: LCSH: Animal communities. | Social behavior in animals. | Animal behavior.
Classification: LCC QL775 .W36 2022 | DDC 591.7/82—dc23
LC record available at https://lccn.loc.gov/2021024308

ISBNs: 9781541600836 (hardcover); 9781541600843 (ebook)

LSC-C

Printing 1, 2021

CONTENTS

INTRODUCTION

Man is by nature a social animal; an individual who is unsocial naturally
and not accidentally is either beneath our notice or more than human.
Society is something that precedes the individual. Anyone who either
cannot lead the common life or is so self-sufficient as not to need to,
and therefore does not partake of society, is either a beast or a god.

—Aristotle

In the rainforest of northern Trinidad, an abandoned house is gradually being reclaimed by nature. Lianas wrap themselves about its
walls, while saplings venture through broken windows and push their
roots through crumbling masonry. Animals, too, have seized the opportunity and have moved in. At the heart of the house, beneath a
sagging staircase, a musty space provides refuge for an animal with a
chilling reputation: the vampire bat. During the heat of the tropical
day, the bats huddle in the cool seclusion of their lair, resting and
gathering strength for the coming hunt. As night falls, they stir into
wakefulness. Their hunger sharpened by the daytime fast, they take
wing to scour the forest in search of blood. They're seeking sleeping
mammals, those that have dropped their guard. Any mammal might

be targeted, from forest deer or peccaries to domestic livestock or even an unwary person.

In a forest clearing, a vampire circles cautiously above a tethered goat. The goat is unaware of the bat's presence, the faint noise of fluttering wings lost amid the many sounds of the Trinidadian night. Stealthily, the bat alights on the ground and scurries in its ungainly fashion toward its victim. It makes an incision in the goat's flank, cutting through the skin into flesh with scalpel-like teeth. As the blood begins to flow, the vampire drinks greedily, consuming as much as a third of its own weight before the meal is finished. Once satiated, it leaves as silently as it arrived—and with its prey none the wiser, despite a wound that continues to flow because of the anticoagulants in the bat's saliva.

Back in the safety of their dilapidated shelter, the bats who have had a fruitful night can begin to digest their meal. But not all the returning hunters have been successful. The large mammals they seek as prey are few and far between, and many of those that can be found are alert to the threat posed by the bats. For these hungry individuals, time is running short: failure to feed on just three consecutive nights can mean starvation and death. Yet this is where the vampire bat's behavior belies its sinister reputation. Should one of its roostmates go without a meal, a well-fed bat will step in. Almost like a parent bird tending its chicks at a nest, the successful hunter provides for its less fortunate companions by regurgitating some of its blood bounty. And the next time tonight's lucky bat goes without, it can count on its companions to return the favor. In their struggle for survival, the bats have each other's backs, a strategy that works well for all in uncertain times.

Cooperation such as this is a hallmark of social animals. Though the extent to which vampire bats engage in each other's welfare is by no means universal, almost all animals that live in groups provide some degree of support to one another. At the most basic level, this might manifest in the form of what's known as social buffering. Essentially, this means that social animals, from tiny krill to humans, gain a clear and measurable benefit simply from being near their own kind and being able to interact with them. They are buoyed by the presence of others, supported by the collective. For our own species, this has never been more important. The recent experience of lockdowns and social distancing prompted by the COVID-19 pandemic disrupted our connections and enforced solitude upon many. Little wonder, then, that a mental health crisis has emerged in the wake of the pandemic. Alongside this, the march of technology is gradually disposing of many of the day-to-day interactions that were once part of normal life. Self-service checkouts, automated tellers, metro ticket machines: all of them replace face-to-face encounters, while headphones lock us out of everyday discourse and the web replaces real-time connections with virtual ones.

The question is: Does this matter? I argue that it does. We humans are intensely social organisms. Our lives are interconnected with networks of friends and loved ones, and each of us plays a role in broader societies that define and shape our patterns of behavior. This social tendency has enabled us to achieve far more than could ever have been possible if we were solitary creatures. Moreover, the far-reaching effects of living alongside one another include everything from the development of spoken language to the ways that we interact with one another in daily life. It even provided the basis for the evolution of the intelligence that is the hallmark of our species. Ultimately, our instinct for cooperation has provided the foundation for human civilization. But this instinct didn't begin with the first people; rather, it was something intrinsic to us, a legacy inherited through our shared ancestry with the animals that we live among.

Countless other animals have adopted sociality to solve the problems that life poses. Living in groups provides the platform for the success of species throughout the animal kingdom. What's more, we can trace direct and important parallels between our own societies and those of the animals with whom we share the planet. These parallels, echoes of our own evolutionary journey, help us to appreciate how sociality shapes our lives so fundamentally. By understanding animals on their own terms, we can understand ourselves, and our societies, so much the better.

Watching animals and studying their behavior has always been my greatest passion. I've spent countless hours at it. Once, as a child, I lay on my front and peered into a tiny stream for so long that a stoat mistook me for a log and decided to come for a drink on the opposite bank, within just inches of my prone figure. When I looked up and came face to face with it, the astonished stoat leapt so high in the air that even its fleas applauded.

But being enraptured by creatures is one thing. Turning it into my life's work seemed impossible. Lacking the confidence to pursue my animal ambitions, and leaving school with a motley assortment of underwhelming qualifications, I drifted into an underwhelming office job. There I stayed for five years, stuck in a rut of my own making. I might have continued much in that vein were it not for the intervention of my manager. It took him a long while to realize the extent of my ineptitude, but once he did, he sacked me.

Forced to find a new direction in life, I pondered what to do next. Could I put my meager skills to use looking after the exhibits at Scarborough Sea Life Centre? It wasn't my dream job, perhaps, but the idea of working with animals, in whatever way, conjured happy memories of afternoons mucking about in rock pools or turning over logs to look for bugs. The Sea Life Centre needed someone to tend to assorted urchins, prawns, and starfish. I reached out to them (the management, not the animals). They reached right back to flick my ears. "You need a biology degree to scrub algae off an old lobster" was the gist of their reply.

At least now I knew what was needed. I enrolled at University of Leeds, where I tried to find meaning in rote learning amino-acid structures. Two years in, just when I was reaching a crisis point in my quest for the degree, I met a kindred spirit in the shape of Jens Krause, one of the academics at Leeds. Here was someone whose curiosity for the living world was as great as my own. Not only that, he was making a wonderful career of studying the behavior of animals.

All of a sudden there was a point to it all. I saw for the first time what I wanted to do. Everything was laid out before me, even if the path wasn't always easy. I like to think that the animals of Scarborough's fourth-busiest attraction wished me well, and have found it in their tiny hearts to forgive me for abandoning them.

The point of these reminiscences isn't just to show you, dear reader, how daunting I found it to be to admit to myself that I wanted to be a scientist. The point is how the purpose of my life became clearer just by virtue of coming into contact with another human being who shared my interests. The realization that you need other people in order to help you understand your own mind and what you want from life is a common experience for many. Unfortunately, what is also common is the human tendency to undervalue this very need, and not to appreciate that community and collaboration are frequently the bedrock of progress and a meaningful existence.

The one aspect of human behavior that has contributed more than anything else to our remarkable success story has been sociability— the ability to live and work alongside one another in groups, to cooperate. It has allowed us to find solutions to problems from prehistory to the modern day, to safeguard ourselves from predators and to hunt our own prey, to share information and learn from one another, to explore the globe and overcome a multitude of challenges. Since the first modern humans appeared in Africa around 300,000 years ago, society has changed and evolved with us. For perhaps the first 290,000 of these years, we lived as hunter-gatherers in small, nomadic bands. Then, as the world emerged from the most recent ice age, the warming climate and human ingenuity ushered in the Neolithic Revolution, and we began to live for the first time in small settlements as agriculturalists. From there, human civilization developed apace, and as it did so, we had other species accompanying us, such as cattle, goats, and dogs, that were all social animals like ourselves.

Modern human society is a mix of culture and relationships, of law and conflict, and is composed of families, communities, cities, and nations. We might imagine that in this regard we stand distinct from other animals. Yet while our society is certainly different in character from theirs, in many ways it is not unique. Many social animals organize themselves in similar ways. Moreover, they were doing so for millions of years before we appeared. Our social instinct, our society, has an ancient lineage, and we have much in common with other social animals. In a world seemingly full of individuals navigating cityscapes and isolation, we arguably need to understand these parallels now more than ever. Why? Because they are a reminder of the fundamentals that shape behavior. Studying the social behavior of animals not only provides insights that are valuable in their own right, but also sheds light on the evolutionary basis of human sociality.

Take language, for example. Communication is an essential facet of living in groups and interacting within a social milieu. The more complex the web of relationships, the more important language becomes. It allows us to navigate our communities, to negotiate, to foster and develop relationships, to teach and instruct. In addition, it has enabled us to assemble into coherent, cooperative teams, from the hunting parties of our ancestors to the organizations and institutions of the modern day. Our cultures, our social behavior and social norms, the rules by which we interact, and the moral frameworks in which we operate developed alongside language. And while human language and culture may not look exactly like any language or culture found among other group-living animals, many animals do communicate with each other, whether it's bees with pheromones or whales with clicks and whistles. By learning something about how other animals live together and communicate, we can understand ourselves a little better.

In another example, think of how we feel about our closest personal friendships. These feelings not only operate at the level of consciousness, but also permeate our physiology, via hormones that cushion us from the worst effects of stress. Those who engage in an active social life and feel close to friends and family members tend to live longer than those who do not. It should come as no surprise, then, to learn that the same may be said for other gregarious creatures. While the assistance that vampire bats offer to their roostmates provides a compelling example, it is the intangible, enduring support garnered simply through interacting with a community that provides the most powerful aid to social animals. Here again, we can appreciate the common links between humanity and the rest of the animal kingdom.

This appreciation has been slow in coming, but scientific research over the past half century has forced a reappraisal of our understanding of animal sociality and cooperation. In recent years, technology has afforded us remarkable insights into the behavior of animals in swarms, schools, flocks, herds, and even our own crowds. These insights have shown us that there are often striking similarities between ourselves and our animal cousins. Simultaneously, they have allowed us to better appreciate the complexity of animals while recasting our own sociality as a fundamental animal impulse. Some balk at such notions, believing humans to be separate and exceptional from other animals. Yet the differences between us and the rest of the animal kingdom is, as Charles Darwin said, of a degree rather than of a kind.

Almost a quarter of a century on from my first fumbling attempts to work out my course in life, I can look back on a dream fulfilled and at a series of adventures. It's been an incredible privilege to study at close hand some of the world's amazing creatures and to grapple with the hows and whys of their social behavior. In the following chapters, I consider a succession of animals, beginning with Antarctic krill and working through to our closest relatives, chimpanzees and bonobos. What all these animals have in common is that they are social.

The word *social* means many different things to many different people, but for the purposes of this book, I define a social animal as one that is drawn to its own kind, that lives and interacts in groups. It's been my life's work to study these interactions between animals: how they relate to one another; how they connive and compete, on the one hand, and unite and cooperate, on the other. In this book I've attempted to distill the wonder that I still feel in the company of animals.

CHAPTER 1

BROWN ALE AND CANNIBALISM

Krill and locusts form Earth's greatest aggregations,
though their motivations differ . . .

THE FROZEN SOUTH

I'm in Hobart, the beautiful capital city of the Australian island state of Tasmania. In front of me in the harbor is a ship, the *Aurora Australis*, Australia's Antarctic flagship. It is a vivid geranium orange, though patches of brown rust show through the paint. You'd struggle to call this an attractive ship, but it is sturdy. It's made countless trips south—to Macquarie Island, situated about halfway between Tasmania and the South Pole, and to Mawson, Casey, and Davis Stations on the Antarctic mainland. The journey to and from Antarctica is not for the fainthearted: traversing the Southern Ocean means entering some of the most inhospitable waters on Earth, a place of terrifying storms and deadly climatic conditions. In this part of the ocean, far from land and shelter, the winds can reach almost 100 miles an hour, well in excess of what feeble landlubbers would categorize as a hurricane. At such times, there's no clear distinction between sea and sky: the furious wind drives the surface waters into a maelstrom of spray, whipping the tops off mountainous waves that can fling a ship around like a toy. Blizzards strike to produce whiteout conditions, and icebergs lurk to claim the unwary.

Happily for me, I can put thoughts of nautical terrors aside. I'm here to visit the Australian Antarctic Division, situated safely on dry land. It's an impressive complex of buildings, decorated with breathtaking images of the frozen south, a part of our planet that few are lucky enough to see firsthand. Outside the entrance there's a triptych of sculpted penguins, arranged as though they were having a chat at the back end of a huge recumbent metal seal, while in the foyer huge pictures capture the ethereal beauty of the Antarctic. Even the food in the cafeteria is themed—you can get burgers that are named for polar scientists. Doubtless, the giants of groundbreaking early expeditions would be thrilled by their commemoration as a snack.

Splendid as all this is, it's nothing compared to the work taking place within. My particular interest is in finger-length crustaceans called Antarctic krill—I want to work out how and why they swarm. It's an important question, because swarming is crucial to krill, and, in turn, krill swarms are crucial to the survival of the entire Southern Ocean ecosystem. Here at the Antarctic Division live one of the only populations of krill outside their natural habitat far to the south.

I'm met at reception by Rob King and So Kawaguchi, two people who've done more than just about anyone else alive to unpick the krill's mysteries. Getting the krill here to Hobart in the first place is no easy matter. They have to be collected at sea and then mollycoddled for weeks on board before the *Aurora* returns to port with its precious cargo. Rob, a genial yet imposing man, described his first, eventful trip to the Antarctic. As he headed south, the weather progressively worsened, until the ship was facing 40-foot waves and vicious winds: a succession of giddying climbs up huge ocean rollers, each followed by a stomach-clenching lurch as the ship surfed down the wave's back. Each time the ship reached the trough, the bow crashed into ocean, and tons of icy seawater flooded the decks and then streamed from the gunwales as the ship staggered into the next climb. Making little progress into the teeth of a storm, the ship was like a boxer pinned on the ropes, taking blow after punishing blow.

Concerned at the damage that was being done, the master was forced into the decision to turn about—a perilous prospect in such seas, because going side-on to barn-sized waves can easily roll a ship and sink it. All aboard knew that, if the worst happened, the prospect of rescue in a storm like this was slim. Even when you're wearing an immersion suit to protect against hypothermia, the sea temperatures here can be deadly. Nevertheless, with the entire crew holding its collective breath, the ship began to edge around. At the mercy of the Southern Ocean, three enormous waves struck in succession, canting it right over to its beams. But the *Aurora* is made of stern stuff, and each time it heaved itself up from the canvas. Once the stern had turned to the waves, the crew, now running with the seas rather than against them, could ride out the storm in safety. Rob described this experience as "highly engaging."

Finally, after weeks at sea, the ship reached the relative calm of Casey Station, an Australian base on the Antarctic mainland, to be welcomed by a small contingent of highly skilled engineers, support staff, and polar scientists, who were awaiting supplies—and, perhaps just as importantly, new people to talk to.

Having arrived at Casey, Rob was itching to get to grips with the creature he has devoted his life to understanding—the Antarctic krill. It was summer, and now that the storm had blown itself out, the conditions were relatively pleasant, with sunshine and temperatures edging above freezing. In front of the station, the bay was more or less free of ice. Rob decided to take to the waters in a small inflatable boat, to see what he could collect in his net. Sitting at the stern, he was happily dipping for specimens when he felt what he later described as "a presence." Turning around, he found himself face to face with a leopard seal that had risen silently from the water and was now looking him right in the eye. Not many animals—people included—are tall enough to look Rob in the eye, even sitting down. But leopard seals can be 10 feet long and weigh half a ton. These are fierce predators, hunters of penguins and seals; they've claimed at least one human life. Who knows what the leopard seal had in mind. But a moment later, it seemed to give Rob a clue, as it opened its huge jaws to give him a view of its formidable, dagger-like teeth, set into a skull the size of a lion's. And then, as if content that the message had been received, the seal slipped back into the water and disappeared. Rob doesn't tend to take the boat out often when he's at Casey now, but when he does, he makes sure not to sit on the side.

After all of that, there was still the return journey to consider, which involved collecting the live krill that sustain the research program back in Hobart. Another journey through capricious seas, interspersed with the exacting task of capturing delicate animals from the frigid waters they call home. And once they were installed in aquaria on board, Rob and So would have to be babysitters—krill are nothing if not demanding. You might wonder why people go to so much trouble for a load of measly shrimp-like creatures. To understand why, you need to see the bigger picture.

MAKING A KRILLING

The leopard seal and a whole host of other large marine predators are drawn to the Southern Ocean to hunt. Whether directly or indirectly, what supports these animals is krill, small but superabundant crustaceans related to prawns. In fact, there are something like eighty-five different species of krill spread across all the world's oceans. But the one that most people think of when they hear the word is the Antarctic krill. For every person alive today, there might be ten thousand of these creatures in the near-freezing southern seas.

Even though each one is only about the size of your little finger, collectively they outweigh us.

Krill are a "keystone species" in the Southern Ocean. This ecological term derives from the crucial role of the keystone at the apex of a stone arch. Take the keystone out, and the arch collapses. So it is for krill in relation to the animals with whom they share their habitat. From fish to squid, from penguins to albatross, and from seals to the great whales, krill are at the top of the menu. Many of these predators have diets that are more than 90 percent krill at certain times of the year. If krill disappeared, they'd take the bulk of Antarctica's most charismatic and important species with them. For the predators, switching their diet to a different prey species is simply not an option; without krill, there would be no Antarctic ecosystem as we know it—no baleen whales, no seals, no penguins, no albatross, and none of the animals that feed on those that eat krill.

Numerous though they are, Antarctic krill are not invulnerable. Twenty years ago, on the other side of the planet, a change in oceanic conditions in the Bering Sea drove the development of a massive algal bloom. Good news for these algae-eating crustaceans? Not at all. It was the wrong kind of algae for the resident Pacific krill, the sister species of the Antarctic krill: they couldn't eat it. Their population crashed, and with them went enormous numbers of seabirds. Salmon failed to show in the rivers, and the emaciated carcasses of whales washed up on shores. The devastating knock-on effects of the slump in Pacific krill foreshadowed what could happen if Antarctic krill were to go the same way.

For now, they are thriving. Drawn together into great aggregations, Antarctic krill can be seen from space when they gather together. A single superswarm might cover more than 100 square miles of ocean, staining vast swaths of the surface waters orangey pink as they cluster in their trillions. Congregating provides krill with some protection from predators and may even help keep them afloat. Since they are heavier than the surrounding water, they start to sink the moment they stop swimming. Yet, by gathering together, they are buoyed by the upwelling currents that result from the countless pulsing limbs of their fellows pushing water downward. The swarm is essentially krill's life-support system.

Although we often think of invertebrates as instinctive creatures, devoid of any but the most basic responses and reactions, krill exhibit a fundamental trait shared by all social animals, including us—they hate to be alone. If they're isolated, they react badly. It's hard to know what panic might look like in an animal that doesn't have a face, as such, but we can measure something akin to it by looking at what is going on inside their bodies. And since krill are largely transparent, it is possible to do this without too much difficulty: we can see their tiny hearts beating. Separated from the multitudes in the swarm, a krill's heartbeat quickens. Krill show a similar response if they detect that whales are around. A raised pulse is a basic sign of stress. Clearly, they prefer company.

Nature documentaries rarely feature krill, but when they do it is as fall guys. We might get only a fleeting glimpse of these small crustaceans, usually portrayed as obliging little floating morsels as they're swallowed by a leviathan. Krill, in other words, are little more than whale food to TV producers. But there's much more to them than this. For one thing, they are far from sanguine about disappearing down a whale's gullet. Despite the numbingly cold waters in which they live, they have surprisingly fast reactions when danger threatens. It takes only around 50 milliseconds for an alarming event in the krill environment to trigger an escape response in them. To put that into context, that's about twice as fast as the reaction of an Olympic sprinter to a starting pistol. The escape response itself is dramatic—in the crucial first second following the detection of a threat, the krill may travel over a meter. Again, compared to a human sprinter, and scaling the krill to human size, that means that the krill would finish the 100-meter race in under two seconds. With a little warning, krill can even elude the cavernous maw of a feeding whale.

In short, catching them isn't as easy as you might think, even for the largest mouths on the planet. Challenging the common notion of whales simply turning up and harvesting the krill, a recent study of humpbacks in the endless days of the Antarctic summer showed the effort they have to put into feeding. The whales lunged at the swarming crustaceans every fifteen seconds or so, minute after minute, hour after hour. With every mouthful, many krill are captured, but still more dart out of the way, leaving the feeding whales short-changed. It's exhausting work for the whales to satisfy their enormous appetites.

But while krill are first-class escapologists, it is their swarming that encourages the whales to focus their attention on them. So why do they collect in such vast numbers? The answer is that the krill are pursued by many different predators and swarming provides an excellent defense against most of them. Any predator that relies—as most do—on picking out their victims one by one faces a kind of sensory overload when confronted by myriad swirling krill.

The little crustaceans have another trick or two up their sleeves as well. According to one account, krill faced with an onrushing predator, such as a fish or a penguin, sometimes spontaneously shed their skins. Anticipating its moment of victory, the predator clamps down on what is now no more than the hollow shell of the krill's body, while the intended victim races to safety. Another oddity is that krill are able to light batteries of bioluminescent cells on their undersides. As yet, no one is certain whether this serves as a means of communication among one another or is used instead to confuse attackers with a pulse of light, or perhaps it breaks up the outline of the krill swarm in the ocean depths when they are being attacked from below. Whatever the reason, the light show certainly adds to the mystery of these engaging little creatures.

Although the interactions between whales and krill are very much in the mold of a predator-prey relationship, it's not entirely a one-way street. To illustrate, we can look at how whaling affected krill. Whalers killed some two million whales in the Southern Ocean in a period of just fifty-five years, between 1915 and 1970. In almost every food web, when you remove a key predator, the prey, freed from

persecution, flourish. But this didn't happen for Antarctic krill. According to some estimates, the krill population declined alongside
the whales. The explanation for this oddity is that being whale food
helps the krill thrive. Whales eat a colossal amount of food—for blue
whales, as much as 4 tons per day—and what goes in must come out.
Whales typically crap near the surface of the ocean. If you've ever lain
awake at night wondering what whale shit looks like, allow me to en-
lighten you: they don't produce a great whale-sized log; it's much more
of a massive, explosive, nuggety cloud of Brown Windsor soup. This
is something I learned as I watched from a boat, with an exquisite
mixture of delight and horror, as a snorkeling colleague of mine became engulfed in one such gargantuan cetacean bum detonation. In
any case, the bits and pieces within the terrible cloud are buoyant and
remain near the surface of the water. Whale turds are full of nutrients, such as iron, phosphorus, and nitrogen, which in turn are manna
for the minuscule plants—phytoplankton—that the krill eat. So the
whales and the krill are bound in an ecological cycle, the success of
one supporting the other.

It's strange what you can find out in the course of research, but it
turns out krill are partial to Newcastle Brown Ale. This isn't quite as
left field as you might imagine. It isn't as though scientists went through
the drinks cabinet to find the krill's tipple of choice. The brown ale was
chosen as an easily available source of dissolved minerals. The point
was to test precisely what the animals found particularly attractive
and, ultimately, how nutrient gradients in the ocean might affect krill
movement patterns. It turned out that krill are strongly attracted to
one nutrient in particular: iron. This is something that dark ales have
in abundance. So delighted were they to be presented with brown ale
that the crapulous crustaceans had to be prized from the pipette that
was being used to add the ale to their tank.

In the wild, krill approach areas with high concentrations of iron-
rich whale crap in anticipation of a bounty of planktonic food. When
whales are active, krill typically spend more time at the water's surface, gorging on the abundant phytoplankton that are fertilized by

the huge mammals' lavatorial habits. Well-fed krill grow faster than poorly fed krill, and thereby have a better chance of reaching an age at which they can breed, even though being at the surface potentially puts them at risk of being eaten by the whales themselves.

KRILL ON OUR SIDE—AND ON OUR MENUS

Aside from their role as the main players in the Southern Ocean ecosystem, there are other, less widely understood reasons to appreciate krill. One has to do with their role in pushing carbon dioxide into the depths of the sea, where it may be locked in place for centuries. Given the role of carbon dioxide in global warming, any reduction in this gas is a bonus for all animals on the planet. The vast numbers of krill are supported by tiny single-celled algae that bloom throughout the Antarctic summer. As they grow, these algal cells suck up carbon dioxide from the water. Krill gather the algae using feeding baskets—specially modified legs that mesh together, which they pull through the water as they swim. They thus take up the carbon in the algae, spit out sticky globs of indigestible matter, and poop out strings of digested algae, which gradually sink. This process, whereby the krill export carbon from surface waters to the abyss, is known as a biological pump. And the abundance of these crustaceans is such that they do this on a massive scale, moving enormous amounts of carbon out of harm's way. It's estimated that Antarctic krill remove as much carbon as is produced by all the households in the United Kingdom every year. Although other algal feeders can pack away carbon in a similar way, they are often far less effective than krill at pushing it into the ocean depths—with the result that the carbon cycles back into the atmosphere.

Very few of us have ever seen an Antarctic krill in the flesh, and at present krill forms only a small part of the human diet. But krill are nutritious, and—even now, so far as we can tell—abundant. Fortunately for them, they're far from a taste sensation. I'm told that to perfectly mimic the experience of eating krill, you need only get a piece of toilet paper and dampen it slightly. Next, put it in the freezer for an hour. Remove and serve. Nevertheless, unlike the toilet paper, they are rich in protein and oils, and their potential both for human consumption and as feed in aquaculture is attracting interest. If there's one thing that, for now, might keep the krill out of our clutches, it's that catching them in large quantities remains a significant challenge. Not only are they typically found in some of the most dangerous seas on Earth, but the fine-mesh nets that are required to catch them clog rapidly, and the animals can be crushed as they are hoisted out of the water. Vacuuming krill out of the sea offers a solution to the net problem, but it doesn't avoid injury to the animals, and that's where the problems start.

Adapted to life in such an icy environment, where their bodies are at the same glacial temperature as the seas around them, krill are reliant on a bizarre internal chemistry. For instance, they possess some of the most powerful and unusual digestive enzymes known in nature. Enzymes are biological catalysts, vastly accelerating processes such as digestion. Our own enzyme processes, and those of most other animals, slow down dramatically as the temperature drops. Krill enzymes, though, have some extreme working conditions to cope with, and consequently, they've evolved to be superpowerful. The amazing characteristics of krill enzymes have recently been harnessed for human medicine, to treat wounds and infections, bedsores, gastrointestinal disorders, and blood clots, to name but a few applications. Scientific advances on this scale are rare. When they do occur, it's amazing how often they emerge in the most unlikely circumstances, from research into what might loftily be regarded as unpromising animals. It's another reason to make sure that we do our utmost not only to safeguard krill but to value all the biological wealth of the planet.

The difficulties involved in fishing for krill are considerable, yet the riches on offer to anyone who can solve these problems remain a powerful incentive. Fishing fleets from several countries, including China, Japan, South Korea, and Norway, are pushing into the Southern Ocean, seeking ways to exploit this massive and essentially ownerless resource. Quotas for fishing are set by international agencies with a view to conservation. But another problem is that no one knows for sure how much krill is out there, and without accurate data, quota setting is something of a guessing game.

Perhaps the greatest potential problem for the krill is that their gregarious nature could be used against them. At certain times of the year, a majority of the krill population may be concentrated in a relatively small number of gigantic swarms, where an intensive, targeted fishing effort could make a killing in every sense of the word. Add into this mix the potentially disastrous effects of global warming, which is reducing the ice sheets that juvenile krill depend on for grazing, and ocean acidification, which, among other things, can prevent krill eggs from hatching, and anyone who cares about the ecosystem of the Southern Ocean has real reasons for concern. To tackle these problems, we need good scientific data to inform our decisions. As Rob King told me, the desire to meet these challenges is what motivates him to go to work each day.

THE KRILL PROGRAM

All of which brings us back to Hobart, and the Australian Antarctic Division. I set off with Rob and So from reception in the direction of the lab for my first meeting with the krill. Strange though it may seem, I'm as excited to see these animals as any of their arguably more charismatic brethren that I've come face to face with over the years. It's true they're not fearsome, like lions, or imposing, like the whales that eat them, but they are extraordinary in their own right. They come from a different world, a place of ice and storms and mystery.

So leads the way. He's a soft-spoken man, not given to grandiosity, yet he's a world leader in his field. He was a key part of the team that first successfully bred krill in captivity at Japan's Port of Nagoya aquarium in the late nineties. He's replicated that feat here in Hobart, making it the second place—and thus far, the only other place—to manage to do it. When you're studying as important an animal as krill, and one that lives in a place as hostile as any on the planet, the only way to gain insight to the mysteries of their life cycle is to breed them in captivity. That way, we achieve understanding of what makes these unique animals tick at close hand.

The krill lab is a modest series of small rooms. There's some high-tech equipment, and there's also quite a bit of jury-rigged kit that's been converted and shaped to its particular, exceptional purpose. Krill have very specialized requirements that are more or less unique in aquaculture, so you have to be creative and learn as you go: there's no instruction manual. As we pass the quarantine area, my eyes are drawn to a series of brightly lit cylinders that stretch floor to ceiling. They glow, each a slightly different shade of green, the perfect decor for a comic-book depiction of a mad scientist's lair. It turns out that this is krill food, cultures of different algae in each cylinder. These get blended together to make a liquid Antarctic salad. Just like us, krill don't do well if they have a limited diet. In their natural habitat, they might eat up to 250 different species of algae. It isn't possible to rep-licate this in captivity, but Rob does his best as the crustaceans' chef de cuisine. Adding to these living cultures, Rob has experimented with products from around the world, and now he has the perfect krill chow—a thick, dark greenish-brown liquid. It smells atrocious to me, though perhaps Rob would make a fortune if he were to mar-ket it to hipsters as a phyto-smoothie. Regardless, the krill apparently greet it with excited, spiraling swimming motions.

And then we go through to meet the krill. There are thousands of them milling around in a collection of huge bowls. The progress of each one through the water is stately—adapting to their polar habi-tat means that life happens at a slower pace. The concepts of up and down don't seem to mean much to them: they swim every which way—on their backs, their sides, their fronts—continually wafting their feeding baskets through the water to extract food. At a sud-den flash of light, though, their insouciance is forgotten as they dart from all corners into a pulsing, defensive ball. Gradually, the cluster splits up again, and the krill disperse. Occasionally, one flickers with a blue-green light, phosphorescing a krill semaphore. Who knows what it means?

The individuals I'm peering at now have been brought from the Antarctic. To keep them happy, the water is held just above freezing by powerful chillers. It doesn't strike you just how cold this is until you put your hands in the water, as I had to do when setting up our filming equipment for my research. Any more than a few seconds of contact with water this cold is surprisingly painful. Getting the right temperatures for the krill is the easy part, however. To successfully breed and raise them in captivity, their needs must be met in full. They need to not only survive but thrive, and they have some very particular requirements. The most challenging of these is water quality. Krill are sensitive to contaminants—not something they need to worry about so much in the pristine waters of their natural habitat, but here it's an ongoing battle. Everything from plastic by-products to metals poses a threat. That means that every part of the system has to be constructed to very exacting, nontoxic requirements. In the early days, when the system was being set up, some unscrupulous contractors tried to cut corners with some of the materials used. The result was a krill wipeout and the loss of eighteen months of precious study time. Rob now watches everything that goes into the krill tanks with eagle eyes. Even if the materials pass his approval, everything must be thoroughly washed in ionized, completely pure water before it can go in with them. There's good reason for this diligence—it's not as though you can pop out to the pet store for more krill. The only resupply is in the krill grounds of the Southern Ocean some 2,000 miles away, and then only during the Antarctic summer.

Having satisfied the housing and feeding needs of these fussy customers, Rob and So moved on to the next challenge—breeding the krill. In the wild, they breed in the Antarctic summer, from January to March. Mood lighting in the aquarium rooms is set to match the wild conditions, with twenty-four-hour daylight in the summer and total darkness in the winter. Based on underwater filming in the Southern Ocean by So and his colleagues, we now have a much better understanding of krill reproductive behavior. Now, while it's fair to say that krill lovemaking is rather different from ours, some of the shenanigans might sound familiar. The ardent male pursues a female and, if he catches up with her, will probe her with his spiky head to find out if she is receptive. If she is, the pair will embrace, which, touchingly, they do face to face. Clasped together, the krill couple ready themselves for the crucial moment when the male wraps himself around the female and presents the lucky lady with a packet of sperm (I like to think he says, "Ta-da!"). Finally, he pushes against the female's underside, possibly to help secure the sperm packet. His job done, he swims off, leaving the female to do the next part on her own. Not that she's what you might call an attentive mother. She produces up to ten thousand eggs, migrates with the other members of the mothers' group to deep water, and . . . drops them. Now she's done, too.

From here on, each egg, with its cargo of a developing embryo, is on its own. The eggs are heavier than water, so they sink slowly downward into the abyss. They might be a mile or two down before they hatch. When they do, the hatchlings—known as *nauplii*—face a journey back upward that, in miniature, is the equal of any of the world's great migrations, a trek of thousands of feet for an animal smaller than a full stop. Why would Antarctic krill parents put their offspring through this? After all, lots of other krill species have eggs that don't sink. One reason may be the presence of the krill swarms themselves. Sinking puts the eggs out of harm's way, out of the reach of the relentless feeding baskets of trillions of older krill.

Whatever the reason, the nauplii are well equipped for their epic swim upward through the icy darkness of the deep ocean. Fortunately for the youngsters, and perhaps to make up for having dropped them off in the first place, their mothers have provided their offspring with a kind of packed lunch in the form of energy reserves that will keep them going for the first month. It's just as well, because the nauplii don't have mouths. As the journey unfolds, they molt and grow, developing mouths and digestive systems. They can't stop—it's crucial that they reach their feeding grounds at the surface before their food supplies are exhausted. As summer gives way to autumn, the surviving young krill arrive at their destination, still less than a tenth of an inch in length, having traveled the equivalent of a marathon (if you scale to human size) every day for a month, all without eating. Their arrival coincides with the return of bitterly cold weather, and the sea ice expands around the Antarctic. But far from spelling death for the long-distance swimmers, the underside of the ice is an inverted savannah to them. They graze algae from the frozen surface like herds of tiny, upside-down wildebeests.

In Hobart, the krill larvae don't have to undertake the same kind of odyssey, but these minuscule and delicate youngsters still demand special care and attention from Rob and So and their coworkers as they grow. It takes up to two years before they reach adulthood. They're in no rush: the conditions in their Tasmanian aquaria are so much to the krill's liking that some of them live there contentedly for years. Krill may live for six years in the wild, and some of those in Hobart stick around for quite a bit longer than this, soaking up the TLC that's lavished upon them.

The opportunity to study krill up close provides enormous scientific benefits. The krill milling around in their icy aquaria in Tasmania are delivering up the secrets of their life cycle and their incredible swarming behavior; they're even allowing us to predict how life will fare in the seas of tomorrow, and much more besides. If we are to conserve the biological riches of our planet, we need to make decisions based on good data, and while krill may not be as emblematic or as cute as a panda, finding out how they tick is the key to preserving the Southern Ocean.

A PLAGUE OF LOCUSTS

As the dominant species on Earth, we have, to a large extent, controlled and marginalized the animals that might do us harm or damage our interests. But one animal has the capacity to devastate human populations on a huge scale, and it is one that we are largely powerless to counter. This isn't some fictional modern megalodon or man-eating tiger; it's a type of grasshopper, the locust. These insects, when clustered together as an insatiable, ceaselessly moving army of billions, are nearly unstoppable and can devastate everything in their path. They spell disaster for people whose lives intersect with their own. They devour every last shred of green, stripping crops, bushes, and trees completely bare of leaves. In the aftermath of a swarm, the countryside is blighted, looking as bare as if wildfire had passed through.

The swarm's arrival is announced by the susurrating, churring sounds of countless wings and the crackling of their jaws on the vegetation. Their numbers can blot out the sun. In the path of the swarm, people desperately try to repel them, setting light to tires, digging ditches, and spraying them with pesticides. But such efforts rarely work. Individually, locusts are vulnerable, but collectively they're relentless—nothing can stop a mega-swarm of them on the move.

In 2004, following heavy, unseasonal rains, a huge outbreak of desert locusts caused misery for the people of northwestern Africa. One swarm, first recorded in Morocco, stretched unbroken over a vast area of land, its width the equivalent of the distance from London to Sheffield, or Washington, DC, to Philadelphia. Within that swarm were ten times as many locusts as there are people on the planet. They laid waste to the countries they passed through, chewing carefully nurtured crops into stalks. As the food ran out, the locusts moved on. The reach of these swarms is colossal. It demonstrates the scale of the problem to learn that a breakaway group from the swarm, a mere 100 million locusts, found landfall on the island of Fuerteventura, over 600 miles from where they had originally set out. Even greater distances have been covered—such as in 1954, when a swarm flew from North Africa to the United Kingdom (in 1988, another swarm crossed the Atlantic, starting from West Africa and ending up in the Caribbean). These unimaginable insect plagues threaten a fifth of the world's landmass, including some of the poorest countries, bringing destruction wherever they go.

It would be bad enough if that was all, yet different parts of the world have their own species of locusts to contend with. In recent times, both Central and South America have been heavily affected by booms in the numbers of their native locusts, while China and India have each suffered periodic and highly damaging outbreaks. In 2010, a swarm of Australian plague locusts infested an area the size of Spain in the agricultural heartland of eastern Australia. Whichever species strikes, the knock-on effects of a locust swarm may go beyond the immediate damage to crops. As the locusts eat themselves into oblivion, their carcasses start to pile up, and the locust bust becomes a boom for other animals, such as rats, which feast on the sudden windfall. Thus one plague begets another.

LEG TICKLING

Locusts have a very different kind of sociality from the Antarctic aggregations described earlier. While swarms of krill are bellwethers for a healthy ecosystem, locust outbreaks are a crisis. If we're to manage locusts and to tackle the blight of their swarming, we need to understand them. That's where science comes in.

The first question is a simple one: Why do locusts swarm? We've made significant progress toward an answer in recent times. Desert locusts exist in two different forms, or phases. In their solitarious phase, where they are hermit-like, avoiding contact with their own kind, locusts are comparatively benign. It's when they enter their gregarious phase that swarming occurs. In a real-life version of the Jekyll and Hyde story, the animals are transformed from quiet, unassuming loners to foot soldiers in a collective eating-machine that is the stuff of nightmares. The two phases are the same species, yet each behaves and looks radically different from the other. So different, in fact, that it wasn't until around a hundred years ago that we worked out that they're the same animal. In the solitarious phase, they are a dull, mottled, khaki-green color, perfect for camouflage. They move slowly and keep a distance from other locusts. Gregarious-phase locusts have a vivid livery of blacks, yellows, and oranges, and they are much more active than solitarious ones. More importantly for swarming, they no longer avoid one another. Quite the reverse—they are drawn together, a necessary precursor to swarming behavior.

As they surrender the stealthy and secretive existence of the solo locust to become pack animals, locusts expose themselves to danger. And to counter the danger, they beef up their defenses. They actively seek out bitter-tasting plants that they eschewed when living the quiet life. Bitterness is a signal of a high pH, the flavor equivalent of a barbed-wire fence. These plants produce alkaline toxins to dissuade leaf munchers from devouring them, but the locusts aren't seeking a toothsome treat. Instead, they stock up on the plant's chemicals, sequestering them for their own defense. Swarming locusts advertise their newly acquired distastefulness via the bright colors of the swarming phase, warning would-be hunters to keep off.

What triggers the switch from their relatively harmless solitary phase to becoming mass marauders in huge social groups? Early investigations seemed to suggest that locusts were more likely to swarm as their populations increased. But why? The answer to this question was provided by scientists armed with some unusual equipment. In 2001, Steve Simpson, now a colleague of mine at the University of Sydney, was investigating whether the transformation from the solitary to the gregarious phase in locusts was caused by physical contact between the animals. Sometimes, real life can be far stranger than fiction. In this case, a major advance in science was achieved through very unusual means. Armed with an artist's paintbrush, Steve and his colleagues diligently and repeatedly tickled specific parts of solitary locusts' bodies for five seconds every minute. The outcome of the experiment was stunning—simply stroking the hind legs of solitary desert locusts periodically over a period of four hours caused them to morph into their marauding, hyper-social form.

Steve's dexterous work with the paintbrush effectively mimicked the conditions that prevail when the insects' larder starts to run low. Desert locusts typically live in harsh arid environments. The supply of food is small, and the best strategy for the locusts is to space themselves out through their habitat and to rely on their camouflage for survival. The arrival of the rains grants a kind of release. The plants respond rapidly, flourishing in the favorable conditions. Making hay while the rain falls, the locusts in turn start to breed. The difficulties begin when drier weather returns. The mass of newly emerged juvenile locusts quickly polish off the abundance of greenery and start to crowd into islands of vegetation. Despite their distinctly antisocial tendencies, the young locusts are forced together by their hunger. The remaining patches of plants shrink under the assault of the locusts' insatiable appetites. They find themselves living at ever closer quarters, bumping into and brushing up against one another. It's this bumping and brushing that Steve was replicating with his targeted tickling. Logic told him that the hind legs of the locusts were a good candidate for treatment because of their size and prominence and the fact that they are covered in sensory hairs. True enough, this was the only site on the locusts that triggered the transformation. Importantly, and unlike other parts of the locust's body, the outside of the hind legs is one area where the locust is less liable to accidentally stroke itself into gregariousness.

The first stage in the change from solitary to gregarious is behavioral; the other changes occur more gradually. The stimulation of the hind legs releases a flood of serotonin to course through the locust's body, and it's this that flips them the full one-eighty from recluse to party animal. The same chemical, serotonin, is of course also active in our own bodies. It's a neurotransmitter linked with reduced aggression and more constructive social behavior. In locusts, the increase in serotonin activates a cascade of physical changes: the Damascene conversion to social behavior is most obvious, but it's followed by a large-scale remodeling of the insect into a fully fledged, insatiable crop-destroyer.

MOVING TOGETHER

The locust, once this transformation has taken place, has a new persona and a rather dapper color scheme. It hangs out with other locusts, but it does not swarm—yet. Something else needs to happen to get the gang moving, and that something is deeply sinister. Unlike the many other grouping animals for whom sociality is a positive thing, locusts are driven by fear. The high densities that trigger the development of the gregarious phase come about through rapid reductions in the available plant food. At this point, the locusts start casting around for other food sources. But there is just one item on the menu that has the perfect balance of nutrients for a locust, and there's an abundance of it: other locusts. The insects now have a potential meal in front of them and a potential cannibal behind. They start to move, forced onward by the threat from those that follow. If one of them stops, the unsentimental locusts behind it may very well treat it as a meal.

As more and more locusts join the party, all of them must march in the same direction. It's no time to be individualistic—falling out of step or changing direction within the surrounding mass of hungry cannibals essentially means volunteering to be a snack. Based on experiments, we know that it's the danger from the rear that drives the locusts onward. When locusts' eyes are shielded to prevent them from seeing the threat behind, or when they can't feel other locusts on their heels, they don't march, but just stay still. Then, as the experimenters say, they get their bums chewed off. Ultimately, the transformation of the locusts into their gregarious alter egos is an emergency measure in response to the risk of starvation, and the movement of large numbers of locusts is a forced march.

Locusts, like krill, aren't particularly hands-on parents. Still, they do at least provide a modicum of help for their youngsters before abandoning them. In the locust's case, when she lays her eggs, the mother uses her recent experiences of life to decide her offspring's destiny. If she's been crowded and jostled amid a swarm of other gregarious locusts, the female will anoint her eggs with a chemical cocktail. This allows her progeny to develop directly into swarmers, without first having to go through all the tiresome leg-rubbing that she perhaps had to endure. The net effect is that, once swarming starts, it takes on its own momentum, and the behavior crosses directly from generation to generation. This is why it's so difficult to break up swarms once they have formed. The transition back to the dull green, solitary, stay-at-home phase gets underway once locusts find themselves alone, but it's a much slower process than the switch to gregariousness, occurring gradually over successive generations. It can take several months.

Even now, in our high-tech digital age, large parts of the globe remain at the mercy of a biblical pest. Our knowledge of the hows and whys of swarming has increased exponentially, but this is yet to be translated into a practicable solution. At present, the most commonly used method to control locusts is pesticides, often sprayed from aircraft once the swarms are already on the march. It probably feels better to try something rather than do nothing, but the effectiveness of this approach is limited in the face of the scale and persistence of some of these outbreaks. It also has major drawbacks: it's expensive financially, and it wipes out some of our insect allies, such as crucial pollinators, alongside the locusts. More sophisticated methods are being tried, such as using the locusts' natural enemies against them. One such enemy is a fungus from the genus *Metarhizium*, which infects, weakens, and kills the locusts without affecting other species. It may be possible one day to target wild locusts by manipulating their serotonin response, so that they don't make the Mogwai-to-Gremlin switch in the first place. Perhaps then the specter of destructive swarms of locusts bringing hardship and misery to millions of the poorest people on the planet will be no more than a memory.

THE LEAST-LOVED ANIMAL?

When I moved to Australia a few years ago, I prepared myself for a series of encounters with a gamut of deadly animals, including crocodiles the size of pickups; spiders that lurked inside of shoes, ready to ambush unwary feet; and legendarily fearsome snakes that could kill you with a sideways glance. If you took yourself to the seaside for a bit of respite from this zoological onslaught, and if the sharks didn't snaffle you, there was a thimble-sized octopus that could cancel you in the blink of an eye.

I'm still here, so I guess I'm either a born survivor or they're assuming—probably accurately—that I would taste bad. In any case, as ever, it's the things you don't prepare for that cause the most consternation. On the first day I ventured onto the streets of my adopted home, I saw a huge cockroach brazenly strolling down the pavement in broad daylight, as though it were window-shopping. I'd never seen one before, but I immediately categorized it as a pestilent, verminous beast. That set off an internal dialogue in my mind. How could I objectively come to such an instant conclusion about an animal? I gave myself a good talking-to. Here I am, a paragon of rationality, a torchbearer for the less well regarded of creatures, yet I was curling my lip at an insect, squinting at it with unbridled contempt. Not for the first time, I was disappointed with myself. But, try as I might, I couldn't reason it out. I was viscerally disgusted by the mere sight of this earwax-colored, hexapodal horror. There's just something repellent about cockroaches, whether it's how they scuttle around at speed, or their hairy legs, or, in the case of the American cockroach, their size. A loathing for roaches is rooted deep within our brains.

Our disgust is well founded. Cockroaches carry pathogens on their cuticles, such as salmonella and *Escherichia coli*, and can potentially contaminate food simply by walking on it. Many bacteria can happily survive the roller-coaster ride through the insect's gut to parachute from its backside with a package of poo and thereafter molder away and taint whatever's nearby. Worse yet is that some of the proteins in cockroaches' bodies are powerful allergens to humans. They can cause asthma, or, in the case of prolonged contact, trigger severe, possibly anaphylactic reactions. And, as anyone who has ever suffered an infestation can tell you, they're tough to eradicate. A good friend of mine lived through a severe outbreak of German cockroaches. These are only small insects, with a rather natty little pale band across their bodies, but they can build up huge populations if left unchecked. Moving into a new apartment, my friend found himself with half a million lodgers who would emerge to party each night and for whom nothing was sacred, including his bed as he slept. He managed to exterminate them in the end, but not before they, in turn, had exterminated his sense of humor. Part of the reason that cockroaches have made something of a comeback in recent times is that cockroach baits, once highly effective at battling these insects in our homes, seem to have lost their edge. There's a suggestion that cockroaches have learned to avoid the sugars that are used to draw them into the baits, and that word has spread through the cockroach community.

Cockroaches are famed for their hardiness. While claims of headless cockroaches surviving to lead a lasting and fulfilling life are wide of the mark, they can go without food for weeks, or simply switch to unusual food sources, such as the glue on the back of a postage stamp. The other reason they're hard to eradicate is the rate at which they can breed. A population started from a single pair of German cockroaches can grow by a factor of a million in a couple of years, given the right conditions. Unfortunately, the conditions we inadvertently provide for them tick every box for the discerning roach. There's warmth, freedom from predators, and an abundance of hiding places and food in our houses. As things stand, it's hard to see any likelihood other than that we will be cohabiting uneasily with cockroaches for the foreseeable future.

Given the fact that I recoil whenever I see one, I'm far from a perfect advocate for cockroaches. Consequently, I'm not going to try to persuade you that they're actually very decent little creatures who are good to their mothers and that we should be more considerate to them. But as a group, they do deserve a fair hearing. For one thing, there are nearly 5,000 species of cockroaches. At most, only around 30 of them cause us any problems. In fact, the other 4,970 species avoid us just as much as we avoid them, and go about their business performing a variety of important ecological roles. Still not convinced? Well, it's a hard sell, I admit. But by understanding these animals better, we can gain insight into how animal societies evolved and learn how we might be able to control them more effectively.

COCKROACH COHABITATION

Many cockroaches are sociable and live in groups. Mixing with others plays a vital role in their development from an early age. Interacting with their own kind as they grow is important for cockroaches, just as it is for other social animals. For example, we know from studies that people who have suffered isolation in their formative years will have difficulty in forming bonds with others later in life. They're also less playful and have much slower language development. Without wishing to draw any kind of direct equivalence between humans and cockroaches, I think it is fair to say that gregarious animals who don't experience a stimulating social environment in early life fail to thrive later on. A cockroach isolated in early life is in fact a tragic figure. It grows more slowly than its peers, and, even as an adult, it finds itself on the margins of its society. Unable to mingle and interact properly, it struggles to join cockroach groups and has an unfulfilling love life. If only they could write, such cockroaches might produce haunting poetry of surpassing beauty and pathos about their existential misery.

Usually, when we see a cockroach, it is on its own, and if you've just turned the light on, it's scrambling for a dark corner where it can hide. This behavior is different from that of most of the animals I describe in this book, many of whom spend much of their lives within an arm's (or a leg's, or a fin's) length of others of their kind. Cockroaches tend to spend their days secreted away in dank places alongside their fellows before going off on solitary feeding expeditions at night. Their social behavior occurs largely out of our sight.

In the seclusion of their daytime hideout, cockroaches gather in distinct groups. They navigate their social world by smell—each individual has its own chemical signature that allows others to identify it. Using their exquisitely tuned olfactory sense, they can distinguish between members of their own group and outsiders, and they can tell their relatives from other, unrelated cockroaches. This is what allows them to organize their communities, sometimes with several generations of the same families living side by side. Beyond social recognition, cockroaches use their chemical capability to learn who's been eating what. In this way, the cockroach group becomes the nerve center of their nightly foraging operations. Knowing this about them might be useful to us—our present pest control solutions either rely on damage limitation, by containing their numbers, or on nuking them with cockroach bombs. Although the bombs certainly work, they use some pretty nasty substances. If we could tap into the cockroaches' chemical language, we could potentially develop far more effective and better-targeted means of pest control, luring them into traps. As we've already seen with krill and locusts, research into sometimes unpromising subjects can provide major benefits.

A group of cockroaches will usually remain in their shelter for as long as there's enough food nearby to sustain them. Unlike some other animal groups, they aren't aggressive to outsiders, so the weary traveling cockroach can find shelter among a group of strangers. Living in groups not only provides the roaches with the social interactions they crave, but also improves their conditions. In the larger gatherings, the huddle of cockroaches raises the temperature and humidity slightly, helping the young grow faster and preventing them from drying out.

If the food supply runs out, or if the shelter becomes too crowded, it may be time for at least some of the group to move on. In this case, the emigrants will seek a new shelter that ticks all the cockroach boxes—darkness, shelter from the elements, nearby food, and, most of all, other cockroaches. Among this group of pilgrims, there will be a mix of personalities. Some of the cockroaches are explorers, ready to shoulder the risks of finding a new home, while others are inherently more cautious. When the explorers locate a dark and cozy bolt-hole, their presence makes it more likely that the other roaches in the group will come and join them. Given a choice between an empty shelter and one that has cockroaches already in it (or at least smells of cockroaches), they will choose the latter every time. In this way, cockroaches attract more cockroaches.

This urge to hunker down in groups is so powerful that it overrides any other information that the cockroaches use to pick their new home. Even if they individually perceive that a shelter is wrong in some critical way—if, for example, it is too bright—they will usually accept the decision made by previous cockroaches to take up residence there and move in themselves.

Despite living in groups, cockroaches are individuals. They lack a clear hierarchy or a leader to tell them what to do. This makes them different from, say, ants, who live in groups with highly structured social behaviors, defined roles for individuals, and a queen who is central to the running of the colony. We know that ants, bees, and wasps all evolved from solitary wasps, but the origins of another insect that forms highly organized social groups, the termite, was long shrouded in mystery. Then, just over ten years ago, a close molecular examination of termites revealed that they are, in fact, a kind of super-social cockroach. So the study of cockroach society leads on to the study of the colonial insects, whose social groups are among the most complex and fascinating of all animals.

CHAPTER 2

HONEY, I FED THE KIDS (AND NOW I'M GOING TO EXPLODE)

Bees, ants, and termites are insect superorganisms.

MAKING A BEELINE

I'm walking through an English wood in spring, reveling in nature. The vivid green of the young leaves is a wonderful antidote to the pervasive gloom of the countryside just a few weeks earlier. On the forest floor, wood anemones and other wildflowers are racing against time—before long the budding trees will form a canopy over them blocking the light. Birds vie for territory, proclaiming their tenancies in song. There's a vibrancy in the air, and I'm just pausing to take it in when a hairy golf ball flies past.

Except it's not a golf ball, but a queen bumblebee. The size of these fantastic, fluffy regal insects always takes me by surprise. The queen, like the other animals and plants in the wood, is on a mission. She has spent the dark days of winter biding her time alone underground. Having all but exhausted her internal larder, she is busy collecting nectar from the early flowers to build her strength before turning her attention to nesting. She might opt for a hole in the ground, or an old mouse nest, or something more upmarket, such as an unused bird box.

In this instance, the queen that just thundered past me darts into a cavity in a tumbledown, drystone wall. Once, this must have been a boundary, built by a farmer long ago. Now it's been overtaken by trees and brambles, a thing of no real consequence until the queen moved in. I stoop for a closer inspection. I can hear her inside, busying herself in the way that bees do, but I don't see her in the darkness of the wall's interior. A moment later she emerges. I watch her make a quick stop at a couple of flowers, then she's back to the wall. It seems she's settled on the site for her nest, putting all her eggs into this crumbling basket. Well, there are worse choices—mossy and broken as it is, it has been here for decades and will doubtless last a little longer. But I can't help thinking how much work is ahead of her. Using her own wax, she will sculpt a pot and fill it with nectar to support herself. She will have to tend her eggs, and she'll no doubt be an exemplary mother, dedicated to the task of nurturing her progeny. This is no mean feat in the brisk springtime of the neighborhood. For the eggs to survive she'll need to keep them warm, 30°C (86°F), ideally. She will achieve this by curling up against her clutch and buzzing her flight muscles to generate heat. Even with a stock of supplies laid in, she will work up a hunger with this effort and rapidly deplete her food. So, every so often, she'll have to bolt from the nest on a lightning mission to pick up supplies from nearby flowers. The moment she leaves the eggs they will begin to cool, and at this delicate stage that can be a major problem. She has just a few precious moments to top up her larder before she must return. She will keep this up for four exhausting days, until at last her attentiveness will be rewarded, and her brood will hatch.

Still, there will be no letup for the queen—she will now have perhaps a dozen hungry mouths to feed. Several days of relentless to-ing and fro-ing between the nest and the woodland garden will follow. She will cram food into her young until they develop sufficiently to begin spinning their silk cocoons. At that point, the queen will finally be able to take a breather. Inside their cocoons, meanwhile, the larvae will transform into adult worker bees, and once they emerge, they will take over the tasks of foraging, cleaning, and providing defense. This is the queen's sole job for the rest of her life: to produce young. She will not leave the nest again. Some say she's a prisoner, but I'm not so sure. She's following the example of her own mother and that of countless queens who came before her; if her efforts pay off, she will have fulfilled her destiny. She has founded her own colony, in her own small way re-creating one of the most amazing of all animal behaviors.

I returned to that same spot several times over the spring and summer. Each time, I was gladdened to see slightly smaller, rather more svelte versions of the queen to-ing and fro-ing from the little hole in the wall. These had to be the kids, so clearly the queen had succeeded: her family had established itself. These bees, along with similar animals, such as ants, wasps, and termites, are sometimes described as social insects. They form tight-knit collectives where each individual contributes to the success of a colony as a whole, in some cases sacrificing their own lives to the greater good. Colonies that might comprise millions of individuals seem to act as though possessed of a single mind; the many become one, so much so that the colony is often considered a *superorganism*. According to this line of thought, the social insect colony is like an organism, but instead of interacting cells, tissues, and organs, the constituent parts are the closely coordinated members of the group. There are other defining characteristics of such colonies, including the restriction of breeding rights to the queen, while a host of workers, the queen's offspring, collaborate to raise their siblings. In some insect societies, you see a number of so-called castes, with certain individuals specializing in particular tasks and acting very differently from their nestmates. The interdependence of the members of these colonies lies at the heart of the astonishing success story of social insects across the world.

COMING OUT IN HIVES

When people imagine bees, they usually think of honey and hives, queens and workers. If so, they're surprised to learn that this is far from the norm. In fact, the majority of the twenty thousand or so species of bees are solitary. They live, forage, and reproduce alone like most other insects. But within this group of fascinating and incredibly diverse creatures, we find just about every kind of social system imaginable, from solo operators to those living in small, loosely organized groups of just a few individuals, to the amazing colonial bees (including the bumblebee and the honeybee) who in surrendering their individuality gain something far greater: a synergy that has no equal in nature.

While living in huge, cooperative groups is clearly a very successful strategy for lots of bees, the question is, How did this arise? In particular, how did it come to pass that any insect would sacrifice her own individuality and chance to breed and instead submit to what might seem the lesser role of raising another's young? Bees provide a unique insight into this question, revealing species at virtually every stage between solitary and entirely social.

Take, for instance, orchid bees, the flying metallic jewels of South and Central America that are attracted, as their name suggests, to orchids. The flowers not only yield nectar and pollen but also provide male bees with a chemical toolkit that they use, like master *parfumiers*, to beguile females. Following a successful mating, the female constructs a cell and lays an egg within. Meanwhile, other females are doing exactly the same thing; in some species, they club together and nest alongside one another. Similarly, in some mining bees, the considerable effort that goes into digging out a nest site is best shared among a group of females who all benefit from the collective effort. What's more, these communal living arrangements present potential nest raiders on the lookout for easy pickings with a difficult challenge—someone's always at home, on guard. So we see solitary animals gathering together to lay their eggs, perhaps representing a first step on the road to colonial life.

Sharing an abode is a step toward community, but if each bee is an independent operator, then they're still a long way from the spectacular social arrangements of honeybees. Carpenter bees come a bit closer. These sometimes live as a twosome: either a pair of sisters, or a mother and daughter. But despite the notion that sociality is based on mutual respect and tolerance, their family relationships can be fraught. One member of the pair is typically dominant and will firmly put her nestmate in her place, eating any eggs she produces. The dominant bee softens this blow by acting as the chief forager, departing the nest to collect food, and leaving the other bee to guard the entrance to the nest and deny entry to predators, scavengers, and other house-hunting carpenter bees. In this arrangement we see a basic version of the complex societies formed by their more social cousins. The right to reproduce is held by only one individual; there is a division of labor; and all the colony members, in this case just the two of them, play a role in protecting and raising the young.

It's hard not to think of the junior partner in this relationship as being disadvantaged by the scheme, so why does she put up with it? It turns out that even though the subordinate bee suffers the indignity of having her eggs eaten, she can benefit through something known as *inclusive fitness*. When we biologists talk about fitness, we don't mean hitting the gym for a workout, but rather, how good an animal is at passing its genes on to the next generation. Most animals gain this kind of fitness by having their own young, but this isn't the only path to success. They can also gain fitness by helping out their relatives, who carry the same genes as themselves. Inclusive fitness simply means fitness in the broadest sense of passing genes on. The carpenter bee helpers gain fitness indirectly, by helping out their kin. And it's not all high-minded philanthropy. Competition for nesting sites can be intense, so the options for the helper may be limited in any case. Besides, if the junior bee can put up with the status quo, she'll get to inherit the nest once the old termagant dies.

These tentative steps toward group living give us some idea of how complex insect societies arose in the first place. It's thought that the path toward social living began with female bees remaining at their nest to help out their sisters or their mothers. As an aside, it is always the females who are social—the males of these species almost never are. But it would be wrong to imagine that such species are inevitably marching toward a goal of cooperation and sociality. In fact, some species have retreated from sociality back to a solitary way of life. Societal living does not suit all animals all the time.

WORKING UP A SWEAT

I'm not always the most sociable animal myself. Many years ago, visiting friends in Maryland, I found myself dragged outside to play softball on a hideously hot and humid afternoon. My preference was to stay wrapped around the air conditioner with my book; instead, I found myself, red-faced with the heat, sulkily trying (and failing) to leather the ball into an adjacent cornfield and so bring the game to an early conclusion. I was saved a prolonged humiliation by the intervention of animals. One moment I was railing internally at the concept of organized fun, the next we were attacked by a relentless horde of small flying insects, heading straight for our eyes. Eventually we took shelter inside, leaving the invaders to go off and be a pain in someone else's eyes.

My rescuers were sweat bees, who supplement their diet by drinking salty moisture from the faces of hot softball players, among others. In hindsight, I should have cried them a little saucerful of saline as a thanks for their efforts, but I was far too focused on pressing my face against the sanity-saving blast of the aircon.

Another upside to these bothersome creatures—at least from the perspective of a biologist—is their fascinating social behavior. Sweat bees keep their options open. Sometimes the females make their own way in life and nest on their own. At other times, or in other places, they form social colonies. The deciding factors that tip the balance in favor of nesting socially are the degree of competition for nest sites and the environmental conditions, which, when friendly, can allow the bees to raise several broods in a single year. A study of a rather charming British sweat bee with natty orange legs, *Halictus rubicundus*, showed that bees from cooler climates keep themselves to themselves. As temperature dictates how much food is available, and how fast the young grow and develop, bees in the north don't have the opportunity to raise several broods in a single year. Farther south, where it's fractionally warmer, the bees can really go to town. They can raise more than one brood during the warmer months, so the first generation to hatch can stay around to help raise the next. Larger families need more food than smaller ones, but with plenty of willing wings available to collect food, and plenty of flowers to supply it, working as a team rewards all of them. In general, nesting in a colony is the best option for these bees if circumstances allow, but their flexible approach to sociality means that sweat bees can adapt their behavior to suit the conditions.

Building a nest and maintaining a colony requires a concerted effort, so it's perhaps not surprising that some species try to game the system by piggybacking on the efforts of others. Cuckoo bees cleverly exploit the cooperative behavior of related species. While social varieties scramble around, selflessly working to supply their colony, cuckoo bees take things easy—they get other bees to feed their young. A female cuckoo bee with an abdomen full of eggs is an insect on a sneaky mission. Her goal is to enter a colony of another species and drop the kids off. Her deceit is helped by the fact that she typically resembles a worker of her target host species, and she smells like one, too. First, she has to get past the guards at the nest entrance. They're on the lookout for just such subterfuge, but with dozens of workers coming and going, the cuckoo, in her stealthy disguise, has a good chance of making it past security. Once inside, she will lay her eggs in the brood cells prepared by the hosts and rely on an army of duped workers to raise her offspring. Her larva will eat not only the pollen inside the cell but the original occupant of the cell, a develop-ing host larva. Or, rather, it will if its mother hasn't helped herself to the larva first. To add insult to injury, when the alien larva emerges from its cell as an adult, it might simply decide to remain in residence within the hosts' nest, parasitizing their efforts like the very worst houseguest.

SHOW ME THE HONEY

Of all the social insects, honeybees are the best known and best loved by humans across the world. There are good reasons for this, one of which is of course honey. Beekeeping, or *apiculture*, has a stagger-ingly long history—archaeological records suggest that people kept bees in earthenware vessels as long as nine thousand years ago in North Africa, and contemporaneous artworks show ancient Egyp-tians keeping bees over four thousand years ago. Honey has some almost miraculous properties that allow it to be kept virtually indefi-nitely without spoiling. This is partly because of the concentration of sugars in honey. The sugars are *hygroscopic*, meaning that they suck in excess moisture, so that it's difficult for microorganisms that cause spoilage to thrive. Honey also contains gluconic acid and hydrogen peroxide, as a result of the bees' work in processing nectar into the end product. Put together, this makes for a substance that is almost completely immune to the attentions of bacteria. Pots of honey re-covered from Egyptian tombs remain perfectly edible despite the thousands of years that have passed since they were first interred. The bacteria-battling properties of honey are one reason why it's been used throughout history as a natural bandage for cuts and burns— it forms a barrier that microorganisms can't bypass—and it is still used today, either smeared straight on the skin or as a coating for dressings.

But of course honey is most sought after as a food—not only by humans, but also by animals such as chimpanzees and honey badgers, who brave the fury of the bees to access the delicious golden, sugary prize within the hives. Honey is a fantastic solution to the problem of how to store energy—that's what makes it so desirable—but the work that goes into it is incredible. It's no wonder that bees are associated with hard work. Each worker bee might pinball around a hundred different flowers each time she leaves the hive. She will collect nectar from these, sucking it in and storing it in her honey stomach, before she regurgitates it back at the hive. Next, a bee production line processes the nectar, repeatedly ingesting and then regurgitating it, each time partially digesting it and reducing the water content until the nectar is fully processed and can be stored in the honeycomb. A hive can produce up to 90 pounds of honey in a single year, which is all the more remarkable when you consider that each bee produces only a fraction of a teaspoon of it in its entire lifetime.

An established hive is a miracle of cooperation and coordination among perhaps up to fifty thousand individual insects. At the heart of the colony, the queen lays eggs continuously—sometimes producing around two thousand of them a day. It is essential to maintain this rate because at the height of summer, when the hive is at its most active, the life span of a worker bee from egg to death is only around two months. The queen is attended by young workers—her daughters—who act as her courtiers. They feed her with royal jelly, a secretion made by nurse bees; clean up after her; and also (strangely, from a human perspective) lick her. A well-licked queen is a clean queen, but more importantly, it allows the workers to collect chemical messages from her that they can then spread as they wander throughout the colony. The effect of the chemical messages is to reassure the colony that it is "queenright": in other words, that all is well with the queen and therefore the group.

A worker bee's life is not only short but also highly regimented. After emerging from her brood cell around three weeks after the queen laid her egg, such a bee graduates through a series of jobs. The first is to clean her brood cell so it can be used again. The queen inspects the cleaned cell, and if it isn't up to scratch, she may insist on the youngster doing it again. Next, the worker bee reports for duty as a nurse, feeding the developing larvae, and sometimes tending the queen. As she gains in seniority, the worker becomes a builder, constructing and repairing brood cells using her own supply of wax, which is produced by glands in her abdominal wall. After a couple of weeks in these junior roles, the workers diversify, taking on responsibility for guarding the hive against invaders, fanning their wings to control the temperature inside the hive, fetching water in hot weather, and even acting as undertakers, depositing dead workers or larvae far from the colony. Finally, a bee may get to do what bees are best known for: forage for pollen and nectar in the world beyond the nest. How long she will remain in this role will depend on the time of year—at the peak of the season, she may last only three more weeks or so. The job takes a heavy toll: essentially, a foraging bee works herself to death for the benefit of her colony.

A STING IN THE TALE

The selflessness of bees is also demonstrated by their fearless kamikaze defense of their nests. Only female bees carry a sting, which makes sense, since it's a modified ovipositor, an egg-laying tube. The sting is barbed, causing it to lodge in the skin of the target. It snags as the bee pulls away and—fatally—tears away her entrails. The chunk left behind includes a pair of pulsing glands that inject venom into the victim's wound. Honeybees are the only bees to have exaggerated barbs on their stings and, as a result, are the only ones that perish after stinging. However, the stings only lodge in the bodies of large animals with thick skin, including mammals like you or me. Against smaller targets that threaten the nest, the sting doesn't usually snag, and the bee survives.

Honeybees aren't spiteful—they use their stings primarily as a means of defending the nest, rather than for preemptive strikes. Even so, these passive and typically harmless creatures can be transformed into hotheads if they believe that the hive is under threat. When a bee stings a perceived enemy, it also releases a pheromone that triggers an aggressive response in nearby comrades, who will join the attack. Cartoon characters pursued by angry bees might jump into water to escape them, but the pheromones left by previous stings are difficult to wash off, so the bees may well remain in the vicinity, where the scent of the pheromone is strongest, until the victim reemerges. Bee stings can be fatal, especially if you're allergic. Even if you're not, a mass stinging can be extremely dangerous.

Few people have ever been subject to such prolonged insect rage as that experienced by Johannes Relleke in 1962. He was walking with his dog in the bush in what was then Rhodesia when something triggered an attack by bees. In a moment, an ordinary stroll became a headlong dash for safety. He ran for the nearby river, chased by the bees, and leapt in with his dog. Relleke tried to keep both himself and his dog submerged, only exposing his face (and the dog's snout) when the need to breathe became desperate. The bees relentlessly tracked him downstream, stinging him whenever they got the opportunity. His bad day got worse when, during his ordeal in the river, an opportunistic crocodile took his dog. Despite the bees' attentions, he somehow pulled through and now holds a Guinness World Record as the person to have survived the greatest number of bee stings—2,443 separate stings were retrieved from his body. The only lasting effect is that Relleke went deaf on one side. This isn't a known side effect, but the reason for it became clear some years later, when Relleke was thrown to the mat in a judo session and a long-dead bee fell out of his ear.

As well as sacrificing themselves for the colony by attacking whomever, or whatever, they perceive to be a threat, bees will isolate themselves if they know they are carrying a parasite. One of the most famous events of Captain Robert Scott's ill-fated expedition to the Antarctic in 1912 was the selfless action of one of the expedition members, Captain Lawrence "Titus" Oates. On the return journey from the pole, in worsening conditions and with supplies running low, Oates was suffering from severe frostbite. Acutely aware that his condition was slowing the progress of the rest of the party, he resolved to take matters into his own hands. According to Scott's diary, which was later recovered, on the morning of the seventeenth of March, as they sheltered in their tent from a blizzard and temperatures of $-40°C$ ($-40°F$), Oates told his colleagues that he was "just going outside and may be some time." He was never seen again. It is unlikely that bees have any concept of their own mortality, yet by isolating themselves when they're afflicted with parasites that might be transmissible to the rest of the colony, they prevent the spread of the infection. And because the bees are unable to survive outside the colony, this represents an Oates-like sacrifice on their part.

Although bees, like other social insects, are known for their devotion to their colony, the harmony within the nest is delicately poised. The threat of anarchy lurks at the margins of their society. The cause of this conflict is the right to breed. Typically, only the queen lays eggs, but across many species, the honeybee included, supposedly sterile workers develop the same capacity. Although they haven't mated, they can produce unfertilized eggs, and these develop into males. The chance to have a son of their own is too much for some to resist, but their actions represent a threat to the established order. As a result, workers police this seditious behavior. If they find an egg laid by any bee other than the queen, they dispense summary justice by eating it. This is pretty effective, but not completely foolproof— around 1 in every 800 males (known as drones) raised by honeybee colonies is the son of a worker. Although this tiny proportion of low-born males hardly sounds like a problem, if one of them manages to mate with a queen, it results in a dramatic increase in the proportion of fertile workers in the colony and the anarchy spreads. The conflict of loyalties between queen and workers in such a scenario can severely undermine the smooth running of the colony, which is precisely why policing it is so important to social stability. Although it mostly consists of eating worker eggs, sometimes harsh reprisals can be dished out. In some species of ants, for example, breeding workers are attacked by their sisters, who bite their legs to immobilize them and may even drag them outside the colony to die.

PASSING ON THE MESSAGE

In his book *Bees: Their Vision, Chemical Senses, and Language*, the Nobel laureate Karl von Frisch wrote, "The bee's life is like a magic well: the more you draw from it, the more it fills with water." Von

Frisch was a pioneer in the burgeoning field of animal behavior, and by devoting his life to the study of bees, he revolutionized our understanding of them. His enthusiasm for his work was legendary. It was the passion that fuels anyone who is truly immersed in a subject they love—the more you find out, the more interesting it becomes.

One of the chief questions that engaged von Frisch's attention was the language of bees—specifically, how workers communicate information about where to find the best flowers. Noticing that foragers always seemed to perform an unusual behavior on their return to the colony, he tried to understand what exactly the forager was attempting to pass on to her nestmates. He published his theories about the dancing bees in 1927 to widespread skepticism. Nowadays, the "waggle dance" of the bees and its meaning is widely known and entirely accepted, although some of von Frisch's contentions were not fully vindicated until 1999, seventeen years after his death.

The waggle dance provides a beautiful example of animal communication. In it, a returning forager dances in a figure of eight. The critical phase of the dance is the waggle run, where the bee moves in a straight line while vigorously shaking its abdomen. The direction that she moves during the waggle run tells her sisters what direction to take when they leave the hive seeking flowers. The number of degrees from vertical translates to the number of degrees from the position of the sun relative to the hive. So if a bee dances in a straight line, that's fifteen degrees to the right (or clockwise) of the vertical line, and she's telling her nestmates that the food source is fifteen degrees to the right of the position of the sun from the hive. And that's not all. The distance that the bee travels when dancing the waggle run describes the distance to the food source. When she gets to the end of a waggle run, she returns to the start to do it all over again, in case some of her sisters didn't get it the first time. If you imagine a rather fat, squat figure eight, the waggle run forms the middle bar, halfway up the figure. The bee traces the loops above and below as she reaches the end of the waggle run and circles back to the start, once taking a clockwise path, and the next time going anticlockwise.

As any halfway decent public speaker knows, there's a lot more to effective communication than just relating dry facts. If you want to get an important message across, you need to do it with feeling. I try to remember this when I'm talking to students, although I draw the line at dancing. Still, bees seem to have a better grasp of the need for a bit of pep than some people. When a forager bee wants to spread the news about a particularly magnificent flower, she dances with extra gusto. This tells her nestmates that she's excited about her find, and her enthusiasm helps to ensure that they will check it out for themselves. She's told them where the flower is and indicated the riches that await them. But if breaking it down like this makes it all sound straightforward, think again—the foragers perform this dance in the total darkness of the hive's interior in the midst of tens of thousands of crowded, milling bees. It'd be like you or me trying to play charades on a platform at King's Cross at the peak of rush hour with the lights off. How do they compensate for these challenges? The answer is that a dancing bee transmits her message by a variety of sensory means, including buzzing at a particularly attractive frequency to get the attention of possible followers and producing chemical cues that tell them about the food source.

In addition, bees that are drawn to the dancer tend to keep in very close contact with her, sometimes using their antennae so they can feel the direction in which she dances. Funnily enough, the dancer can sometimes get rather possessive about her news. If another bee is dancing, too, she might bump into it to make it stop, thereby ensuring that everyone's paying attention to her. Thus the diva dancer passes on her knowledge, and the followers understand where they need to go, navigating according to the position of the sun. But what if they get outside the hive and the sun is hidden by clouds? No problem—the bees can see polarized light, which means they can determine where the sun is even if they can't see it directly. The accuracy of the information provided by the dance varies, but it is better for long-distance foraging sites than for nearby ones—which makes sense, as the farther you need to go, the better the directions need to be.

DECIDING WHERE TO LIVE

Trapped in yet another academic meeting, trying to make it through a whirlpool of circular arguments, I'm watching the clock and wondering why it is that such gatherings seem to find it so hard to make decisions. Not only this, but in my experience the larger the meeting, the slower the resolution. You might reason that humans are perfectly equipped for the task of decision-making: we can discuss the relative merits of various alternatives and then take a vote on what to do. Yet so often, the sticking points are so sticky we get stuck. Like us, group-living animals are frequently compelled to make decisions and to resolve conflicts over differences in preferred courses of action. Unlike us (or unlike academics at least), they're amazingly good at sifting through information and reaching a consensus.

How do they manage this when they can't conduct a debate and typically don't vote? Looking for an answer to this question, researchers have examined the remarkable process by which swarming bees choose a new home.

The sight of thousands upon thousands of bees clustering noisily in a tree is a sign of a major upheaval in the bees' lives. There are a number of triggers that might cause this swarming behavior. It might happen when the colony outgrows its hive, or it may occur when the queen is old. She typically lives much longer than her offspring, but as she ages, the chemical messages she provides to the colony decrease, to the point where the workers are prompted to begin rearing a new queen. They select one of the embryos and provide her with a prolonged diet of a unique substance known as royal jelly, which they produce from glands on their heads. In addition, they withhold from her the more common fare offered to other larvae. The royal jelly transforms her development and sets her on the road to becoming a queen. With two queens now in situ, the colony faces a split. Swarming divides an existing colony in two: the original one will rebuild under the new queen, while around half the workers leave with the old queen to establish a new base elsewhere.

It's a race against time for the leavers. The only food provisions they take with them are what they can carry in their tiny crops, and this won't last long. When we see them, perhaps hanging from a tree branch, they're waiting for the decision to move on. Some of the bees, acting as scouts, will have left the swarm to locate a new home. They report back with the information they've gathered and present the options by, again, dancing. If a scout finds a particularly good site for a new nest, she will dance enthusiastically on her return, waggling furiously and repeating the dance, potentially over hundreds of circuits. If the site is less desirable, she will be more subdued in her dance and will perform fewer repeats.

The energetic dancing of an enthusiastic scout convinces other scouts to follow her directions and check out the site. If they, too, are enthusiastic, then even more scouts will go take a look. The original scout will keep flying from the swarm to the proposed site and back. Each time she returns, she will dance, but her dancing the second, third, and subsequent times will be shorter and less dramatic. After all, if the scout was just as enthusiastic each time she returned, she might recruit more scouts every time, and the swarm would get stuck in a never-ending cycle. That first scout could be wrong about the site, or just too easily pleased.

Instead, the swarm relies on corroboration of the first scout's information and positive feedback. During this process there may well be a number of small clusters of scouts, each dancing to advertise a potential site. Just as with the foraging dance, the dancers may ram or headbutt bees from other factions. The aim is to stop their opponents from recruiting for a rival site, and it calls to mind the worst behavior of political opponents in human elections. Gradually one side will win over more recruits and drive down the enthusiasm of its rivals. Once a side has recruited a critical number of scouts—something like fifteen of them—the decision is made. How long this takes depends on the number of different options available and how good they are, but it's possible for a swarm of thousands to reach their decision in just a few hours. The final phase begins when the scouts move through the swarm, passing on the message that it's time to warm up their flight muscles to get ready for a mass departure. Once they're all hot to trot, the whole swarm makes a beeline for their new home.

TERMOPOLIS

Years ago, as a child, I sat on the sofa and watched, transfixed, as a nature documentary on termites unfolded before me. I can't overstate the effect that Joan and Alan Root's seminal *Mysterious Castles of Clay* had on me, especially when it takes us inside the huge, architectural masterpiece that is a termite mound, to see for ourselves an insect civilization. It was one of those indelible childhood moments. I can still remember the otherworldly music that played as the camera miraculously teleported me into an immense termite fortress to witness the extraordinary lives of these minute insects.

Everything about this documentary was hypnotic to me. I was no longer the little boy who thought insects were simple and unremarkable. They were now a life-form that displayed sentience and collaboration, representing an alternative plane of existence that enriched my understanding of the world. My younger self's once casual disregard for insects is very much the norm. It's hard enough for people to take a closer look at things that aren't in their immediate field of vision, especially when you consider all the other distractions life can throw at us. Yet, although earthy termites may not hold as much attraction as their soaring and vibrant cousins, the bees, they're just as deserving of Karl von Frisch's label of being a "magic well" of knowledge.

Termites are sometimes grouped together in people's minds with ants, bees, and wasps. But while they have much in common with these hymenopterans, they stand distinct from them as a lineage, and are more closely related to cockroaches. Of the three thousand or so species of termites so far described to science, the great majority are small (usually less than half an inch in length), blind, and soft-bodied. Unlike many of the ants, with which they are so often confused, they are vegetarians, with a particular fondness for dead or

decomposing wood. In 2011, one enterprising gang of termite robbers even broke into a bank in India and ate their way through ten
million rupees worth of paper money. As the bankers found out, you
underestimate these animals at your peril.

Termites are typically found in warmer climes, and where they
do occur, they are staggeringly successful colonists—South Africa's
Kruger National Park alone is thought to be home to over a million separate termite cities. Similar densities occur in the Serengeti
and Masai Mara, which is where I first saw termites in the flesh,
so to speak. If asked to think of the Masai Mara, your mind's eye
may conjure a tableau of long grass, flat-topped savannah trees, and
free-roaming megafauna—a landscape in which termites do not feature. Yet the sheer abundance of termites I saw there made for one of
the most spectacular natural phenomena I've ever witnessed. It was
March, and the ground, the air, the bushes, and the vehicles were
thick with winged insects. Everywhere.

Termites are unaccustomed to flight, and with four long, ungainly
wings they're not the most accomplished aeronauts. Their strategy is
based not on agility but on numbers—countless termites erupting
from their subterranean nests, inundating the landscape. For a few
brief minutes, each of these termites leaves the underground confines
of the colony to soar. This is their nuptial flight, a termite carnival
where the goal is to meet their significant insect other. Amid this
organic tsunami I noticed that the local fauna wasn't slow to exploit
the glut, gorging themselves on the termites to the point where many
seemed barely able to move. Dozens of gluttonous birds staggered
around the acacia scrub, too fat to fly. It soon became apparent that
my traveling companions, John and Joseph, both from Nairobi, knew
a little about the culinary delights of termites. I was encouraged to
try one myself.

"They have a pineappley flavor," John told me. I noticed he wasn't
eating.

"No—more like carrots," said Joseph, before adding, "That's what
my mum told me. I wouldn't eat one."

Part of me was with Joseph on this one, though equally, I hate to miss an experience. So I grabbed one and ate it before I could overthink it. It was crunchy and not terrible, but that's the best I could say. Who knows, maybe I chose the wrong wine to go with it? Still, they are certainly sought after by people in many parts. Despite my depredations and the rather larger inroads made by various local mammals, reptiles, and birds, the supply of termites seemed virtually infinite.

Between the impromptu snacking and the apocalyptic insect swarm coating the ground and stuttering around inexpertly in the air, it was easy to forget that this horde was formed not for the convenience of diners, but for the purpose of mating. Unfortunately for the termites flying around my head, successful suitors would form a minority. For those lucky couples that succeed in pairing up on their nuptial flight, the next task is to shed their wings and get underground as quickly as possible, out of the way of their myriad predators. Amid the soil, the pair will construct a royal chamber and then mate. In most cases, they will remain together until death does them part. This can be a very long time—these regal termites can live for decades. Soon after mating, the new queen will begin to lay eggs. When these hatch, the youngsters will devote themselves to the service of their parents, tending and feeding them and building the colony around them. In time, their mother grows prodigiously large. Compared to her rice-grain-sized offspring, she might be the size of your middle finger. She is, in effect, immobilized by her bulk; her body becomes distended. Her translucent skin shows a churning mass of fats and ovaries—she's a giant, living egg factory. And what a factory—she produces an egg every few seconds for the rest of her life and keeps her attendant offspring busy with the never-ending task of transporting their sibling eggs away to be hatched.

SOCIAL SPECIALISTS

One of the defining aspects of social insects is their so-called caste system, which describes how individuals within the colony take on diverse, specialized roles and often look totally different from one another. The surprising thing about this is that these individuals have the same parents and a very similar genetic code. While human siblings often bear close resemblance to one another, the castes of a social insect colony can look like separate species. At the outset of the termite colony, the offspring will develop into the most common type, workers. These fulfill the roles of building, cleaning, foraging, and rearing their younger brothers and sisters. But for all their scrupulous fetching, carrying, making, and mending, workers aren't the most effective defenders. Soon enough, a successful colony will attract the attention of rivals or predators. If these are to be kept at bay, the colony needs to call on its soldiers for defense. During the early life of a small proportion of the colony's young, a combination of environmental and social conditions causes a developmental switch in their bodies to be flicked, changing the destiny of those youngsters to grow into soldiers as adults.

Depending on the species, soldiers may be armed with huge, powerful jaws, supported by outsize heads containing a battery of muscles to work these weapons. Some guard the entrances to the colony, permitting entry only to their own colony's workers. Others accompany the workers on foraging expeditions, acting as a protection detail as the workers collect food out in the open. Yet another kind of soldier will use its large head to block the passageways within the colony if it should come under attack, apparently willing to give up its life to prevent raiders from accessing the vulnerable queen, king, and workers within. A still more astounding form of soldier termite, the *nasute* soldier, develops a long spine from its head that makes it look like a termite tank. When threatened, these soldiers shoot noxious, irritating, sticky chemicals through this tube to subdue the colony's enemies.

ANT WARS

It's just as well that termites are capable of defending themselves—a termite colony represents a tempting target for many predators. Ants, the insect cousins of the termites, raid termite colonies in search of sustenance for their own colonies. The resultant hostilities between ants and termites are ceaseless and enduring; they've been playing out for millions of years. Any campaign relies on intelligence on both sides. The ants scout for prey, hoping to pick up the scent of termite foraging parties. It's not simply a matter of finding their targets—the ants gauge the strength of the colony, and thus the potential riches on offer. The termites, in turn, have their own form of espionage, monitoring the tiny vibrations produced by columns of raiding ants in the area. The termites conceal their own presence in case the ants are listening, muffling their sounds by tiptoeing around.

If the ants launch an attack, the termites raise an alarm by sounding the drums. The fact that they don't have drums is no impediment, as they innovate by banging their heads against the walls of the mound. Tiny though the sound may be, the soldiers are alert to it and respond by massing to the most vulnerable parts of their fortress. The oldest termite soldiers go to the front lines in this battle. Despite their greater experience, they aren't more effective at defense; rather, their seniority means they are more expendable and so can be sacrificed first. The stakes for both sides are high, and consequently, the fighting is vicious; casualties mount rapidly. Injuries needn't be fatal. In the heat of battle, an injured Matabele ant of sub-Saharan Africa will release a pheromone, a chemical cry for help, that causes its fellow ants to pick it up and carry it back to its own colony, where it may recover to fight another day.

The ants are a formidable enemy. If they fight their way past the cordon of termite soldiers, the entire colony of thousands, or even millions, of termites may be in peril. The largest ants, the majors, engage the termite soldiers, while the smaller members of the ant army stream past the combatants to the interior of the mound. The worker termites are far less fearsome than their soldier brothers or sisters, yet they throw themselves into defense, biting the raiders and clinging to the ant's legs.

Once inside, the ants will push on toward the heart of the nest. Desperate fights occur along the arterial routes of the colony, and chaos reigns as these ancient enemies battle for survival. As a precaution, the worker termites seal the royal chamber with mud, which will harden and thus protect the queen and her consort from incursions. But meanwhile, the battle rages, and the termite soldiers augment their biting, slashing, and spraying with less conventional tactics. According to one account, this includes firing frass at their enemies. It's hard to say what this achieves—perhaps it recruits more termites to the attack through chemical cues, or being pelted with poo affects the ants' morale. A more dramatic and perhaps more effective defense is provided by exploding termites. In some species, the oldest working termites turn themselves into suicide bombers, and actually blow up when raiding ants bite them. They can do this because of a pouch of blue, copper-based crystals they carry around on their backs. When an ant bites such a termite, such that the pouch ruptures and the crystals mix with saliva, the chemical reaction ends in a mini explosion that showers the attackers with noxious chemicals.

Inevitably, the toll will be high on both sides. Even if the termites manage to repel the ants, their colony will be weakened and more susceptible than before to follow-up attacks. But in any case, the ants' goal isn't necessarily the extinction of the termite colony. They may even raid colonies in rotation, in order to maintain a series of viable termite nests that can be attacked at a later date. A column of retreating ants will carry off the spoils of war—the bodies of thousands of the vanquished defenders—back to their own nest as food.

MASTER BUILDERS

Driving through Dubai recently, I craned my head upward to take in the gravity-defying scale of the Burj Khalifa, the tip of its spire almost disappearing in the dust storm that was blowing in from the desert. The skyscraper, about 460 times my own height, is a triumph of human engineering, of materials, power, planning, and hard work. But then compare that to the largest termite mounds. Some species build edifices up to 30 feet tall. For a typical worker termite, that's something like a thousand times its own size. And even though these incredible structures are built predominantly out of the surrounding mud, they can endure for centuries, or, in at least one case, millennia.

From a human perspective, it might be natural to assume that the termites build their mounds or repair them according to some master plan, and that each termite has an overview of what it is trying to achieve. Termites are blind, however, and individually, each one is operating using a very simple internal program, and very likely has no idea of any kind of blueprint for a mound or the entire finished product. How can an animal build something without knowing (or needing to know) what it is building, especially on such a scale? The answer lies in a phenomenon that animal behaviorists call *self-organization*, which describes how small parts of a system can interact without any centralized control to produce a larger pattern or structure. Beyond animal behavior, the beautiful symmetry of snowflakes is one example among many of this idea. A bunch of water molecules clearly don't have a plan in mind when they get together to make a snowflake. Instead, it's just the way the molecules interact with each other that shapes the crystal.

In the case of termite builders, each worker perceives a change in the conditions inside the mound. For instance, an increase in airflow might mean there's a hole in the mound. The termites' response to noticing the airflow is first to collect a particle of moist earth and mix it to a muddy paste with saliva; then it follows the draft to find the hole. When it arrives at the construction site, it deposits its little ball of mud next to the ball of mud that a previous builder laid down, then it returns for more supplies. The process continues, with workers delivering thousands of tiny muddy balls until the airflow stabilizes. This same simple sequence—pick up mud, make a mud ball, put the mud ball next to another mud ball—is enough, given sufficient time and a massive number of termites, to build some of the largest animal-made structures on Earth.

More extraordinary even than the scale of a termite mound are the arrangements within. Chambers connected by termite highways can be found at the base of the mound and belowground, while only sparse numbers of workers on maintenance detail populate the towering insect cathedral above. The mound itself is a marvel of design. Its chambers and passageways regulate ventilation and airflow, letting just the right amount of air in and out. Gangs of workers—like sailors on a clipper ship trimming the sails to the weather conditions—constantly tinker with these passageways to maintain the conditions their nestmates need below. The sun provides the power to drive the air movement, but the termites' smart design makes it all possible. During the day, the outer walls of the mound heat up, causing hot air to rise up the mound's sides and cooler air to be sucked down through the top. At night, as the outer walls lose heat, the reverse happens. On a daily cycle, the mound inhales and exhales like a lung.

The African termites that build these giant mounds have reason to pay close attention to conditions inside their colony—they have an important guest to consider. Actually, it's perhaps more like a partner than a guest, because the termites couldn't thrive without it, and vice versa. This partner is a fungus, and it's the missing link in the termites' success story. Much of the nutritional value of the vegetation that the termites bring back to their nest is locked up in the tough cellulose that plants use to build their cells. The termites can't digest this, but the fungus can. This explains why the termites go to such lengths to make the fungus feel at home. They build special structures known as fungus gardens for this purpose and tend their growing crops with every bit of the dedication of human farmers.

If you were to cut open a termite mound, you'd see these fungus gardens at around ground level. They have complex, whorled surfaces, like corals or sponges, which provide the largest possible area for the fungus to grow and work its magic on the cellulose. Foragers returning from outside keep the fungus gardens going by bringing in a gutful of chewed-up vegetable matter to daub on them. The result—the wonderfully named *pseudo-feces*, a delightful stew of leaves, grass, and spores—feeds the growing fungus, which the termites eat in turn. Other workers monitor the humidity: If the fungus gets too dry, it dies. Too wet, and competing fungi take over and ruin the crop. The termites bring in water or manage ventilation to maintain just the right conditions. From our perspective as animals, it's tempting to see this partnership as one where the termites use the fungus—but the reverse might actually be more accurate: the fungus manipulates the termites into building a perfect, protected environment that provides for all its needs. Either way, the relationship works beautifully on both sides.

In his book *The Soul of the White Ant*, the naturalist Eugène Marais saw in termite mounds a kind of chimera, a single organism of many parts. He suggested that different components of the mound were analogous to our own organs—the outer mound was like skin, the fungus gardens like a stomach, the breathing tower like lungs, and the queen served as the reproductive organs. The whole structure might seem as immutable and unchanging as a skyscraper, yet it continually adapts, repairs, and renews itself like a body. It is made up of the collective efforts of vast numbers and generations of minuscule termite builders. Could these workers be like the blood cells of the body, carrying nutrients around and neutralizing invaders?

Or perhaps in the way that their genius emerges from their combination, termites are more like the brain. Our brains are composed of billions of single cells, called *neurons*. One human neuron on its own has only a very limited ability, but interconnected neurons are responsible for both great art and great science. Similarly, a single, isolated termite is insignificant. Yet a colony of termites working in concert vastly exceeds the sum of its parts. Their spectacular organization makes these tiny insects one of the greatest triumphs of nature.

KILLER ANTS

Filmmakers over the past century or so have repeatedly delved into the natural world to find subjects that are guaranteed to scare audiences. Think of *Jaws*, *Jurassic Park*, or *Piranha*. The idea of humans as prey is a compelling one, jangling nerves by undermining our conceit that we're masters of our destiny. When I wasn't watching nature documentaries as a child or looking for animals in filthy places, I might well be found peeking from behind the sofa while remorseless and insatiable creatures pursued and snaffled some unwary actor. So it was with *Ants!* or *Legion of Fire: Killer Ants!* While the writers might be ridiculed for their plots, their imagining of ants as master hunters is sound—ants are among the most efficient and deadly predators in the animal kingdom.

The likely inspiration for such cinematic horror stories are army ants. This term is loosely applied to a range of different ant species that form huge colonies and also have two other things in common: their lack of permanent nests, and their tendency to embark on huge foraging expeditions. In some species, the ants may form mobile columns stretching the length of a football field.

The main reason for their nomadic behavior is their insatiable appetites—they may capture and eat as many as half a million prey animals every day. If the army ants were to make a permanent home, they would rapidly exhaust the supply of victims in the vicinity, and so, with potentially millions of mouths to feed, the colony must keep moving.

The scale and ferocity of raiding columns of army ants is the stuff of legends. Although they primarily eat other invertebrates, such as grasshoppers, cockroaches, and other social insects, there are tales of them taking larger prey, particularly in the case of African army ants (sometimes known as driver ants), whose colonies may be up to twenty million strong. A mid-nineteenth century account of the American naturalist Thomas Savage describes army ants in what is now Liberia overwhelming pythons, pigs, birds, and even a monkey, while the French explorer Paul Du Chaillu related tales of people accused of witchcraft being staked out on the ground so that the ants could serve as slow but inexorable executioners.

Army ants are opportunists. It seems unlikely that they would target larger vertebrates on a regular basis. For one thing, unless it is injured or constrained, a large animal can easily escape from army ants—their columns typically move at a speed of only around 65 feet an hour. Nonetheless, army ants can subdue prey much larger than themselves by dint of their vicious bite and sheer force of numbers. Large spiders and scorpions may be overwhelmed by a foraging raid of them, their own powerful weapons of no use against myriad biting ants. Overwhelmed, the victim is quickly and efficiently dismembered before members of the column carry it away in pieces.

Some species of army ants target the nests of other social insects, such as wasps. There is little the wasps can do in their defense, other than to flee with as many of their young as they can carry, while the ants mass to destroy what is left behind. As the raiding column moves along, threatened animals frantically break cover to seek safety. This flushing of prey explains the close association that many other species maintain with the species of army ants known as *Eciton burchellii*. Amazingly, over three hundred different species are directly reliant on these army ants for their own livings, the largest number known to associate with any species on Earth. Antbirds pick off insects as they flee from the raiders, while parasitic flies home in on cockroaches driven out of hiding. There are ballsy species of beetles that mimic the ants and live side by side with them.

Army ants are creatures of routine. They spend around two weeks on the march and a similar amount of time encamped in one spot. This pattern is driven by successive waves of the colony's young. When babies hatch, the army turns into a hunting machine to satisfy their voracious appetites. They march by day, with hordes of workers forming the center of the column and fierce soldiers protecting the flanks. The queen accompanies the army, surrounded by an entourage of workers. Specialized long-legged workers act as porters, carrying the precious larvae clasped to their undersides. So the army moves along, raiding and killing as it goes. As dusk falls, the ants form an extraordinary structure known as a bivouac. This is essentially a nest made by the living bodies of the ants themselves—hundreds of thousands of them meshed together into a cluster that can be three feet across. Inside, the queen and thousands of larvae are safely tucked away—few creatures would be foolhardy enough to take on such a ball of malice.

This nomadic phase comes to an end when the larvae begin to construct their cocoons to embark on their transformation into adults. Since the army no longer has to feed the larvae, it can take something of a breather. The daily onslaught of marching ceases, and the colony makes camp. The priority now is to cram food into the queen, whose body swells dramatically in preparation for laying perhaps up to three hundred thousand eggs in just four or five days. The hatching of those eggs coincides with the emergence of the previous cohort from their cocoons, giving the signal for the army to mobilize once more, and a new cycle begins.

Every three years or so, a phenomenal event takes place when the *E. burchellii* colony splits in two. In the tropics, splitting happens at the beginning of the dry season, when the colony has some half a million members. The queen produces eggs, as at each previous cycle of her reign, but this time the hatchlings are reproductive ants, with ambitions of becoming future queens themselves or of mating with a queen.

For all that the language suggests feudal, inherited authority, it's the workers who hold the power. They will decide who will be queen and who will be her mate. They have to choose well—as fearsome as they are, army ants suffer huge losses during their raids, and a healthy queen capable of producing plenty of eggs is essential for the colony to thrive. The splitting ceremony and the election of a new queen is a masterpiece of organization. Only around six new queens will be produced, and most of these will not succeed to the throne. As if sensing this, the virgin queens produce powerful pheromones that serve to recruit a retinue of supporters among the workers even while they themselves are only larvae. The loyalty that these regal larvae manage to inspire means that fights can break out between their rival followers within the bivouac. While they are larvae, the queens are nourished and protected, but once they emerge as adults from their cocoons they face the decision that will determine their fate.

The colony forms two separate columns facing in opposite directions, joined now only by the bivouac at the center. Within, the would-be queens prepare for judgment. One virgin queen will attempt to make her way along a column, accompanied by her worker attendants, while other workers hold back those awaiting their turns. If the procession of the young pretender goes well, the workers will accept her as their new queen. If not, she will be cast aside and abandoned. The process continues until each of the two columns has selected its queen. Perhaps the old queen will keep her role and lead one of the columns, but there are no guarantees. Queen army ants live for around six years—but it will be another three years before the next generation of queens will arise. In that time, she will need to produce millions of eggs and walk colossal distances with the raiding army columns. The workers need to be satisfied that their queen has the stamina for this. If they decide she does not, they will select a new, more vigorous monarch and cast their old queen aside, and she will face abandonment and, inevitably, death. Once the selection is complete, the workers in each of the two columns form a bivouac around their new ruler. Gradually, the two columns break apart, each heading off in a different direction, and the single old colony becomes two.

The males remain in their cocoons a little longer than their sisters, the virgin queens. They're carried off by workers when the colony splits, but when they finally emerge, they won't hang around. Males have wings, and their mission is to fly off to find new colonies, and new queens with whom to mate. Like the queens, they're substantially larger than the workers—so much so that in Africa they were once thought to be an entirely different kind of animal, gaining the name "sausage fly." Essentially, they're giant flying gonads, massive mating machines that, in addition to a lot of sperm, also carry around chemical inducements to impress new queens and their colonies when they find them.

For humans, the arduous task of winning over an object of affection can pale in comparison to that of attaining the approval of a potential mate's family or friends. Spare a thought, then, for these male ants, who, as they arrive, seeking to mate with the queen, must run the gauntlet of aggressive workers and persuade this skeptical audience—using chemical messages—that they're fit partners. If the workers agree, a consort will be brought to the queen, in some cases having first had his wings torn off, and presented to her for mating. It makes sense for the workers to be choosy—they will, after all, be raising his offspring. But the male, alas, doesn't get to see this. He's a one-trick pony: after he mates, he expires in what we might hope is a postcoital haze.

THE WISDOM OF ANTS

In the tropics, where army ants reside, ants may account for more than a quarter of the total biomass of all the animals found there. But it isn't just in the tropics that ants play such an enormous role: ants are everywhere, and often in huge numbers. It has been calculated that at any one time there are something like one hundred trillion ants alive on our planet. Put another way, that's not far short of fifteen thousand ants for every person alive today (although anyone who's ever had a picnic in Australia might think this is an underestimate). You can find them on virtually every speck of land outside of Antarctica, nesting, thriving, and multiplying. In some cases, they form vast supercolonies—networks of interconnected nests spanning enormous distances. The Argentine ant, for example, has become an invasive species, establishing supercolonies around the world. The largest of these stretches from Portugal to northwestern Italy, spanning some 2,500 miles.

Ants owe much of their success to their sociality. Like termites—but unlike their closer relatives, bees and wasps—all species of ants are social. If you see an ant on its own, it's either on a short-term scouting mission or terribly lost. Or it might be a member of one of the many species of insects that, perhaps jealous of the ants' success, pretend to be ants. As we've already seen, the combined efforts of a collaboration of many individuals is a potent recipe for success. Another part of the ants' winning formula is their ability to adapt in order to exploit opportunities. There are ants that feed on other animals, ants that feed on plants, and ants that aren't fussy and eat pretty much anything they come across, right up to and including engine oil.

But their adaptability shows up most clearly in how they solve the problems of their day-to-day lives. The army ant columns described earlier act as a high-speed food capture-and-delivery service for their developing young. They go to extraordinary lengths to maintain the efficiency of this service. Specialized workers even fill in potholes in the route—using their bodies. Ants carrying supplies walk over the tops of these fellow workers rather than through a series of ruts in the trail; thus the workers smooth out the route so the column can move faster as a whole. Similarly, it can often be difficult in patchy terrain for tiny creatures to travel in a straight line. Gaps between the branches of a shrub, for example, might force wide detours. The army ants' ingenious solution to this is to build bridges of their bodies to span the gaps. On either side of what to them is a canyon, the ants cluster and begin to stack themselves one on top of another, like circus acrobats making a human pyramid. Eventually, the two sides meet, and the ants clasp together, making a pathway and allowing their foraging coworkers to move rapidly through. Once all have crossed, the living bridge breaks up, and the insect engineers move on to their next assignment.

In another example, fire ants on floodplains are at risk of inundation in their underground nests if there is a sudden deluge. So, when floods strike, they build floating rafts made of the bodies of their nestmates. By linking together, they create a kind of living waterproof fabric that maintains and supports the colony and preserves the lives of vulnerable youngsters, who get to sit on top of the raft. And they can maintain these rafts for weeks on end, if necessary. Flooding may even be beneficial, because the water carries them to new areas to colonize. Outside their nests, the ants are more exposed than usual, but that only leads them to beef up their defenses, which makes them much more venomous than they were in their old nesting state.

When they eventually wash ashore, the fire ants produce another incredible structure, which serves as stopgap until a permanent home can be found. Clustering around the stem of a plant, they form a kind of Eiffel Tower that may be dozens of ants tall. The load at the bottom is obviously the greatest, so more ants are recruited there, with progressively fewer farther up. The result is a tent-like structure. The bodies of the ants on the outside provide a waterproof layer to repel raindrops, keeping the individuals on the inside dry. Like the termite builders described earlier, individual ants have only the tiniest brains, and almost certainly no conception of the overall goal of building a raft or a tent. But that's no impediment if they're preprogrammed to follow some simple general rules. Moreover, these rules can change according to the situation, allowing them to make different structures at different times.

All animals have to make decisions in their day-to-day lives. Sometimes these decisions are connected into a sequence, so that each choice has knock-on effects for the others. To give an example in a human context, think about a delivery network: drivers and the companies they work for need to find the most efficient routes to deliver products to lots of consumers, each waiting impatiently for their goodies to arrive. Although solutions to this kind of problem might seem simple, in fact they're anything but. The most famous example is the Traveling Salesman Problem.

Imagine a salesperson, or a delivery driver, who has to visit ten different customers, each in a different place. Starting from the depot, he or she wants to find the best route to take in order to visit each customer. For just those ten customers, there are 1.8 million or so possible different routes! This same problem doesn't just afflict delivery companies trying to work out the best routes; similar calculations are needed in the optimization of manufacturing processes— for example, in drilling circuit boards, or in designing a warehouse, or providing electricity to users. Getting it wrong can be expensive and time-consuming. Humans are actually pretty good at finding solutions to the Traveling Salesman Problem, so long as the network is fairly small. But the performance of ant colonies in solving analogous problems is arguably even more impressive.

The Traveling Salesman Problem is relevant to many species of ants because, just like delivery drivers, they move out from a central depot (their nest) to forage at a variety of different food sites. The important thing is to build a network that delivers food back to the nest in the most efficient possible way. And there's a complicating factor—the food isn't always in the same place. New foraging opportunities appear, and others are exhausted, so the ants have to design different networks all the time. They're astoundingly good at this.

Last summer, my youngest son was eating an ice cream cone on the balcony of our apartment on a hot summer's day. As he turned it to tackle a stray drip, the unthinkable happened—the ice cream made a break for freedom and fell with a splat to the floor. While I was inside cheering him up with a consolation lolly from the freezer, an ant found the ice cream. By the time I got back to the balcony, the puddle of ice cream was encircled by hundreds more ants. Then I saw a thin column of ants leading from the ice cream to a large flowerpot. Word had got out of an unexpected bonanza among the ants who had an apartment on the balcony of my apartment.

The secret to both their rapid mass recruitment and their ability to develop foraging networks lies in the chemical trails the ants lay down to guide one another. When an ant finds food, it will collect a portion of that food and then head back to the nest. As it travels, it pauses periodically to lay down a little dot of pheromone on the path, a chemical beacon to guide other ants. In turn, new recruits to the path add their own pheromone trails in support of the pioneer's trail if they, too, approve of the food. It's the ant equivalent of a "like" on social media, positive feedback that quickly generates an ant superhighway. But what about when the food starts to run out? The pheromones are volatile, meaning that they evaporate quickly. This is a valuable property, because it means that, to remain active, the trail needs to be topped up with additional pheromones constantly. When the time comes that fewer ants are able to collect food and take it back to the nest, then the trail will start to fade away, preventing later ants from following a strong trail to nowhere.

Although a pheromone trail is a great system for linking a colony and its food source, on its own it isn't enough to solve the problem of how to forage efficiently. That's because it lacks flexibility—if all the ants have to do is follow a trail, they could get locked into one particular behavior. A good example of this was provided by the American naturalist William Beebe, who on an expedition to South America in 1921 watched a column of army ants that had somehow become trapped into walking in a giant circle. The circle was over 300 feet across, and the ants marched around and around for two days. Many members of the column died before the cycle was broken.

But events like this are rare. In practice, ants are usually able to solve problems like finding the best route through a network because not all of them behave in the same way. While most ants may simply conform, following the pheromone trail, others are much less predictable. This makes for the kind of flexibility and innovation that are so important in problem-solving. Rather than following the established trail, some maverick ants will wander off and explore different areas. In doing so, they may open up a new and exciting food source.

Often there's more than one way to get from A to B. Finding the shortest path is extremely important to the ants' success, because it allows them to improve the efficiency of the trail network. But when they're first faced with two alternative routes, ants choose them pretty much at random. Say two ants reach a carelessly discarded toy on their path from their nest to some dropped ice cream. If each ant takes a different route around the toy, the one on the shorter path will reach the sloppy, sweet mess first. It'll then retrace its steps and head back on the return journey before the other ant has even reached the food. All the while, it will be depositing trail pheromone, the chemical call-out for more to follow. The shorter length of the more direct path builds up a stronger concentration of pheromone, encouraging further recruits. Meanwhile, the pheromone trail on the longer path is less concentrated, and so less attractive. Because the pheromones evaporate, progressively fewer ants will choose the longer path until it gets forgotten in favor of the shortest, most efficient route.

Lots of experiments have been performed to test the ability of ants to find the best path to a food source. We know that they can make some remarkably precise distinctions between paths, choosing the most efficient route. These tests have been excellent for learning just how ants manage to tackle simple problems, but they don't particularly tax these six-legged experimental subjects. They leave an open question: Are ants merely competent route finders, or are they master navigators? A former colleague of mine at Sydney, Chris Reid, decided to push them to the limit with an ingenious and devilishly difficult test. As many studies had already done, he presented a colony of ants with a maze that ran from their nest to a food source. The challenge came in the complexity of the maze, which offered 32,768 potential routes from nest to food. Of these, only two were ideal, in that they provided the shortest journey for the ants. The ants solved the maze inside an hour.

At the time when Chris did the experiment, the received wisdom was that although ants were good at forming these foraging trails in the first place, they didn't fare so well when conditions changed. Naturally, this amounted to a challenge. Of course, it's hard to know whether ants can get frustrated, but Chris did his best to find out. He changed the structure of the maze after they had first solved it, so that they would need to adapt their behavior to solve it again. And again, they rose to the challenge, impressively solving the maze a second time.

Tests like this rely on a mix of innovation and positive feedback. They have given rise to algorithms called *ant colony systems*, or *ant colony optimization methods*, which can be used in computer science to tackle problems in the human world such as the Traveling Salesman Problem. These optimization methods employ virtual ants using the same rules and processes as their real-life counterparts to provide solutions to all kinds of problems, from routing city traffic to setting up the best university timetable for hundreds of concurrent courses, designing antennas and circuit boards, and predicting soil drainage patterns, to name just a few.

STRANGE ASSOCIATIONS

Some of the most fascinating behaviors of ants can be seen in their interactions with other ants and even with completely different species. In the natural world, just as in human lives, being successful can encourage hangers-on and con artists. For instance, if you happen to be a small and delicious insect, and ants—or anyone else—might want to eat you, or at least kill you, then it might be helpful to be able to trick the ants into thinking you're neither a meal nor a threat. And if you've mastered that trick, then being in the middle of an ant colony could be an exceptionally safe place to be. Ant crickets haven't just pinched their common name from their hosts—they get free protection, lodgings, and food from them, too. Entering a colony for the first time, the cricket faces aggression from the ants. At first, it keeps out of their way simply by being quick on its feet, but it can't run forever. If it wants to make a success of this, it needs to blend in, and it does so in a very peculiar way—it copies the way the ants walk. In the darkness of the nest, doing impressions like this helps the undercover cricket avoid detection. If it is transplanted into the nest of a different species of ant with a different walking pattern by a curious scientist, it shows its adaptability and artistry by rapidly adopting the new gait. As time goes by, it gradually picks up the colony's smell, which is the primary means by which ants recognize one another. After that, it can simply pretend to be one of the girls (workers are all female).

All this is very well, but you've got to eat. The ant cricket has this covered. An important part of the success of many social insects is that they feed one another, a practice known formally as *trophallaxis*. Workers store food in their crops and will regurgitate small amounts of it at the request of other colony members in liquefied form. An ant requests food by using her antennae to tap and stroke the antennae and head of the ant that it wants food from. This sharing and spreading of food is vital to the success of ant colonies, not only because it ensures that resources are divided up among the workforce, but also because it causes colony pheromones to be passed among the ants, helping to bind them to the colony's common purpose. Ant crickets are wise to this and have developed the ability to dupe ants into providing food for them in this way—a quick tap on an ant's head, and it gives up food like a vending machine.

SLAVE-MAKERS

Ants have exceptional teamwork capabilities, but they can also employ some intensely dastardly strategies. Different ant species, and even different colonies of the same species, are often extremely hostile to one another. But in terms of out-and-out exploitation, few can compete with slave-maker ants. Amazon ants are specialists in this practice. Equipped with fearsome, sickle-shaped mandibles, they are usually more than a match for their victims, ants of the genus *Formica*. Just as *Formica* ants are not made of Formica, Amazon ants are not from the Amazon: they hail from the United States. A newly mated queen Amazon ant is a creature on a mission. She must find a colony, take it over, and enslave its workers. Normally, an invader attempting to enter any ant colony would be cut to pieces in seconds, regardless of its impressive mandibles, but the queen Amazon ant stands a good chance.

One reason for this is that Amazon ant queens have very little natural smell to them. Since ants recognize one another by smell, this helps her to go under the chemical radar of the hosts. Another reason is that the queen Amazon ant has some ingenious chemical weapons. She can release a pheromone to calm the aggression of the soldiers and workers of her target colony, which buys her time to enter the nest and find her counterpart, the queen of the *Formicas*. Even at this stage, however, her work isn't done. To be accepted by the invaded colony's workers, the Amazon queen must cloak herself in the *Formica* queen's odor.

Horrifically, she gets this from the *Formica* queen in the process of butchering her, licking her victim to take up her chemical disguise even as she bites, stabs, and slashes her to death. Some colonies of *Formica* are home to multiple queens, so the Amazon ruthlessly seeks out and destroys each one. The recognition system of the *Formica* ants is now compromised; the workers accept the Amazon as their own queen. Newly enthroned, she settles in and lays eggs, which the *Formica* ants, now slaves, dutifully raise for her.

Amazon ants have no workers in the usual sense of the word—they do not forage or raise their own young. They rely on their enslaved hosts to do those jobs. But with no *Formica* queen to fill the nurseries with the next generation of slaves, the slave-makers are in charge of an ever-dwindling workforce. If they're to maintain a healthy colony, the Amazon ants must find more victims to press into their service. Scouts therefore set out on reconnaissance, looking for *Formica* colonies to raid. When they find one, they hotfoot it back to the nest with their news. The Amazons mobilize rapidly, recruiting up to three thousand ants for the attack. Astonishingly, the slaves sometimes accompany their masters on these raids and end up fighting their kindred ants. And the attack, when it comes, is ferocious. Under the onslaught, the *Formica* ants generally flee, abandoning their nest to the raiders. Only in the largest *Formica* nests is the fight ever prolonged, but even then, few can withstand an attack from the Amazons.

Rather than relying solely on strength of numbers, other species of slave-maker ants have developed a suite of secret weapons. Some deploy *propaganda substances*, chemicals that instill panic in the defenders, or even cause them to turn on one another. In other slave-maker species, raiders use a kind of Harry Potter–style cloak of invisibility, except in this case, the camouflage is chemical. The raided nest seems not to realize that invaders are in their midst. So confident are the attackers that the raiding party may be as small as just four ants. But whichever means they use, the end result is usually the same: the slave-makers get what they came for—the young of the nest they have attacked, whom they now transport back to their own nest. Over time, a colony of Amazons may kidnap thousands of

Formica larvae. As they continue their development into adults, the kidnapped young imprint on the slave-maker colony. They come to perceive its familiar smells as home, and to consider the slave-makers with whom they are raised as their sisters.

For readers who are naturally inclined to support the underdogs—or underants—in this story, there is some good news. If colonies of ants begin to suspect that slave-makers are in the area, they become wary and increasingly hostile toward foreign ants. Sometimes they move their colony as a precaution, saving themselves from the depredations of the slave-makers. But if the worst happens, and a colony of ants becomes enslaved, what then? In most cases, it's game over, but not always. Sometimes an underground resistance movement develops, and the slaves rebel. This resistance focuses around the nursery. Slave workers have the job of rearing and tending the young. Within the nursery there may be young from their own colony, laid by their own queen before she was killed by the slave-maker queen, alongside those captured from other nests and brought back to the colony, as well as those of their masters, the slave-makers. This is a dangerous time for the very young slave-makers—their lives are in the hands of the slaves. To survive, the young slave-makers are reliant on smelling just like the young slaves they are with. Their disguise is good, but not perfect. If the nursery workers detect the minute differences between their own smell and the whiff given off by the slave-maker young, they will kill them. It's a fascinating arms race—the pressure is on for the worker colonies to evolve ever more complex chemical recognition cues that allow them to differentiate between slaves and slave-makers, while the slave-makers have to keep pace by evolving increasingly sophisticated scent-based mimicry of their hosts.

FARMER ANTS

Cunning as all this trickery is, we can find a counterpoint in a different kind of ant relationship, one that arguably provides one of the most amazing arrangements in the natural world. Sap-sucking insects, including aphids and leafhoppers, are the bane of gardeners and farmers alike. They stab down on the vein of a leaf with their sharp mouthparts and drink the lifeblood of the plant, its sap. Just as the human body relies on blood pressure to deliver oxygen to its organs, plants rely on pressure to deliver the sap, and the nutrients it contains, to their various parts. When a sap-sucker sticks its mouthparts into the plant, the pressure does most of the job of driving the sap into the insect. So much so, in fact, that the insect often has to shed the excess liquid that it taps.

An aphid usually only partially digests this liquid, which is known as honeydew, as it passes through its body, so it is full of nutrients. It takes ten thousand aphids around an hour to produce a teaspoon of honeydew. That may not sound like much, but with enough sap-suckers and enough host plants it can be worthwhile even for people to harvest it. It has been part of the diet of Aboriginal Australians for thousands of years, as well as that of peoples of the Middle East—it's thought that the "manna from Heaven" in the Old Testament may have been honeydew. Although, as its name suggests, this substance is sugary, it also contains proteins, vitamins, and minerals.

What's good for us is often good for ants. Some species of ants collect honeydew from beneath the plants where the aphids are at work, while others take a more ingenious approach. In a way that brings to mind the dairy farmers in human agriculture, species such as wood ants tend a "herd" of aphids as they feed on their plant. Different types of worker ants have different specialties, with some allocated to guarding the aphids, others to milking them, and still others to transporting the honeydew. The milkers' job in particular is extraordinary—they stroke and caress their aphids to encourage them to produce honeydew. Some ants even build shelters around their aphid flocks to protect them from the weather.

The ants can collect a staggering amount of food in this way—one estimate is that a colony of ants can harvest around half a ton of honeydew in a single year. As aphids grow and mature, they ultimately develop wings, enabling them to fly off. It would obviously cause problems for the ants if their livestock were to disappear, but the ants prevent this from happening, either by clipping the aphids' wings or by using chemicals that retard the development of the winged form of the aphids. The ants can even produce chemicals to prevent their livestock from wandering too much. In winter, some ants take aphid eggs into their own colonies before putting the young out to pasture in the spring. And if the plants that the aphids feed upon begin to fail, the ants may carry their flock to a new plant. All in all, the diligence of the livestock-farming ants is remarkable. You might say it's akin to human agriculturalists, but in truth it's the other way around. Ants were doing this millions of years before we caught on.

Just like the domesticated animals that inhabit our farms and houses, aphids have undergone major behavioral changes under the ants' care. Some species of aphids have coexisted with ants for countless generations and have come to rely on their protection. In that time, some have lost part of their wildness. In particular, "domesticated" aphids are less adept than their "undomesticated" peers at jumping to escape predators, and they seem to invest less effort in producing the waxy coatings for their bodies that play a role in discouraging predators. Ultimately, these changes mean that aphids rely ever more on the ants for their protection. The ants aren't sentimental about their livestock, though. If a flock of sap-suckers gets too large and produces more honeydew than is needed, the ants simply eat the excess animals. Worse yet, if an alternative source of food turns up, some ant species eat the whole flock.

OURSELVES IN OTHERS

Social insects are among the most fascinating animals on Earth. They're so different from us, yet the societies that they form have obvious parallels with our own. Like us, they're agriculturalists and builders. They adapt the world to their own needs. They defend their patch and they specialize in different roles. Other than us, they are the only animals that form structured colonies whose numbers can run into the millions. They also share some of the less attractive aspects of our nature, including the exploitation of slaves. There's another surprising resemblance, too. We have an image of social insects as hard workers, but just as in our societies, there can be major differences in the work ethic of ant colony members. In rock ants, for example, as few as 3 percent of the workers constantly strive and push for the greater good, and about 25 percent never seem to bother working at all. The remainder work sometimes and chillax at others.

Social insects are a perfect case of how something can become ever more fascinating the more you know about it. And they're hugely important to us. Of the hundred or so species of crops that are used extensively to feed the human population, around seventy rely on honeybees for their pollination. Without bees, we would be in real trouble. And although their lives are less intertwined with ours, ants and wasps play an enormous role in pest control. So next time you reach for a can of bug spray, think again. We might not always get along together perfectly, but we owe social insects our respect.

CHAPTER 3

FROM DITCHES TO DECISIONS

Schools of fish are capable of making sophisticated choices.

DITCH THE FUN

I'm clutching a net in a stinking ditch, in a field, in the nearest thing to the middle of nowhere in England. It's a cold November evening and a fine drizzle is falling. I've been progressing up this ditch inch by inch for nine hours, hunting for fish. I'm thigh-deep in water, though everything from the knees down is thick mud. Muscles I didn't know I had are hurting, and having a sense of feeling in my feet is a distant memory, even though I'm wearing a pair of insulated waders. There is scarcely any part of me that isn't covered in filthy goop. The same goes for my colleague, Mike Webster, standing several hundred feet away in a different part of the ditch.

Now daylight is fading and it's time to get out and head back to the hotel, where our clinging, eggy-methane, eau-de-ditch parfum will no doubt clear the bar. No matter—we've earned a pint after this. But then in the half-light I spot the approach of an old man with a genial, careworn face, walking an arthritic dog. He won't have seen me—because I'm in the ditch, all but the top half of my head is concealed from his view. I'm concerned. If I emerge before him suddenly like a swamp monster, this oldster's poor, kindly old heart might give out.

I need to make him aware of my presence, especially since it looks like his old dog isn't quite on the ball. So I strike up a loud, tuneless whistle and gradually haul myself out of the ditch before plastering a smile on my mud-spattered face, trying to make myself look non-threatening, and say: "Nice day for it!"

It's a good thing that I took these precautions. He freezes in mid-step with a look of mingled astonishment and disgust before recovering sufficiently. "It is if you're fucking mental," he says. Slightly discombobulated, I call after him: "Have a lovely evening." He responds with a few more earthy chunterings.

When I tell people that I research animal behavior, those who don't write me off as a wastrel often picture a glamorous series of wildlife encounters in exotic places. No one yet has said, "Animal behavior? Then you must be familiar with the ditches of Lincolnshire!" It's true that I have seen wonderful things in wonderful places, but there's plenty to be learned from field sites that to an outsider lack both charismatic animals and visual appeal. The result of my grimy peregrinations in the ditches with Mike was that we found out how populations of our study species, the three-spined stickleback, organize themselves. Their own personal smell comes from what they eat and where they live, just like when you eat garlic or asparagus, or live above a chip shop. Tiny differences in the conditions they experience, even over a distance of just a few feet, can give them a very specific smell. And they prefer to hang out with others that smell just like themselves.

Our sense of smell is relatively poor, so we don't use this sense to the same extent in shaping our relationships as many other social animals do. A human analogy for the way that the fish use smell to distinguish between trusty locals and nefarious outlanders would be the accents and dialects people use in spoken dialogue. Where I grew up, in the north of England, I can easily tell what county, or city, someone is from, and possibly even what valley they're from, just by the way they speak. The ditch sticklebacks do something similar with smell, and they prefer to mix with a specific group of individuals and to live in the area they're familiar with. If you move them to a different location, they very rapidly find their way back home.

BACK TO SCHOOL

I've always been interested in animals, seeking them out in ponds, among vegetation, or under logs. But I remember one day in particular as a child going to Aysgarth Falls in the Yorkshire Dales. Here in summer, the river Ure eases its way through mature woodland, presenting visitors with a glorious picture of the British countryside. Its more tempestuous winter persona has carved out a series of small waterfalls for the river's summer version. These are the perfect places for a hot day, to let the water stream over your back, or to stand behind a fall and pretend, as I did that day, that I was a fugitive hiding from my enemies.

I had a new toy to try out, a diver's facemask. The first time I fastened it on and put my head under the water, I was met with a sight that surpassed any aquarium I'd ever seen: a huge, shimmering shoal of hundreds of fish—minnows—moving through the roots of a tree and among the water plants, lit by dappled sunlight shining through the leaves overhead. I was mesmerized by them. In my excitement, I shrieked to my dad, who was standing nearby on the bank. In true British style, he was mortified that I was making a scene, but I didn't care; I was thrilled. I floated in that river, letting the fish coast around me for hours.

Since then, I have snorkeled in rivers throughout the north, and it's been glorious every time. I've hardly ever seen anyone else do this, perhaps because it's cold, but every time, it's been the match for anything I've ever seen in more exotic places, in reefs and seas around the world. The underwater scene is a natural wonder that goes unnoticed by the vast majority of people, a rich treasure just beyond their sight.

I'm far from the only person to be fascinated by a shoal of fish, of course, or a flock of birds, or any large gathering of animals. There's something arresting about seeing them congregate in large numbers, like some mighty but benevolent army, and then to see the mass transmogrify magically into different shapes and symmetries, moving and turning in unison. The awe that I felt when I first saw my minnow shoal is a feeling common to many who see collective animal behavior in the wild. If it didn't set my course in life, it certainly made me clear about what I wanted to work on when eventually I became a biologist. I wanted to understand animal groups, and fish gave me a great way of doing that.

When the car company Nissan was developing its first generation of self-driving cars, it turned, as developers often do, to the natural world for inspiration. Anyone who has watched animal collectives, such as shoals and flocks, may notice that the animals don't collide. In fact, their movements appear to be closely choreographed, so that, across the group, the animals move as though responding to some invisible conductor. This collision avoidance was something that Nissan was keen to replicate. Over the past few years, close study, particularly of fish shoals, has revealed exactly how massed groups of animals produce the balletic movements that so entrance us.

First, it is important to point out that there is no invisible conductor shaping and dictating their movements. Each individual animal is responding to a simple series of rules. The basic description of the rules, in order of importance, is: if I am too close to my nearest neighbor, move away from it; if I am too far from my nearest neighbor, move toward it; and, if I am the right distance from my nearest neighbor, copy what it does. So the distance between any pair of neighboring animals in the group dictates how they respond. They fine-tune the gaps between themselves until it is a kind of Goldilocks-style "just right," and then copy each other. When Nissan built its first prototype cars, its engineers fitted them with sensors to allow them to do just this. Sure enough, the little robotic cars performed like a swarm of animals. In this way, they could move far more efficiently than human drivers could manage.

These three rules are enough to get animals (and robots) milling about in one place without colliding, which is fine as far as it goes. But they don't help animals respond to the environment they live in—the world outside the group. And if danger threatens, they have to respond in a coordinated way. One of the reasons that so many animals form into groups is that it offers some protection. A predator approaching a group of animals suffers what's called the *confusion effect*. Presented with large numbers of animals, it experiences sensory overload. This is one reason for the findings of multiple studies showing that, as prey group size grows, predator success shrinks. What the hunter needs to do is pick a single victim, but how can it do this when presented with such a bewildering mass of animals? Random lunges at such groups in the hope of getting lucky and snagging a victim almost never work. So the predator often tries to isolate an animal, by chasing and harrying the group. Some predators on reefs, such as jacks, have a smart strategy for this: chasing a shoal of smaller fish against a rock outcrop, to force them to break up into smaller groups. Another trick is to find something that makes one of the animals stand out. In that case, the predator can lock on to its victim and, very often, catch it. This trick also has a name: the *oddity effect*.

Nearly fifty years ago, when ethical controls on animal research were far less stringent than they are today, some researchers tested the oddity effect by putting white paint on the horns of a small proportion of wildebeests in a herd. Sure enough, the odd-looking wildebeests were quickly taken by predators. So looking the same as everyone else in the group is clearly important as a form of self-defense. Indeed, another defensive tactic for animals under attack is to coordinate their behavior with others. They pay special attention to their near neighbors, moving and turning as they do. The maneuvers they perform often happen so quickly that the human eye—and the eyes of the predator—can't keep up.

SPREADING THE WORD

The way that information spreads across animal groups has been likened to how British Royal Navy ships, spread out in a long line off Cape Trafalgar in 1805, signaled to one another by hoisting flags. In this way, they passed a message along to Admiral Nelson, aboard HMS *Victory*, informing him that the Spanish and French fleets had left Cádiz and were preparing for battle. As one animal on the edge of the group detects a predator and responds by a sudden move, its neighbors react by copying it, then *their* neighbors copy them in turn, and so on. This isn't information in the sense that we perhaps think of it—an explicit message. Instead, we mean information in its broadest sense: data that indicates a change in the status quo.

Regardless of the semantics, the information can spread across the group like wildfire, far faster than any one animal can travel. There is a problem with information, though: it can be wrong. If the whole group responded every time one of its members twitched, they'd all find themselves in a state of physical and nervous exhaustion. Groups have neat solutions to this. First, they are likely to respond more strongly to neighbors who make sudden turns or rapid accelerations than to ones who make smaller, more laid-back changes—because the sudden, fast moves tend to indicate something more important. And, second, as the information spreads outward from its source, from one animal to the next, the strength of the response with each exchange is slightly decreased until gradually it peters out. Unless, that is, more information comes along to reinforce the initial message.

But again, animals must travel around their environment rather than just milling about and dodging predators. What if some members of the group have someplace they want to get to? Animals track environmental gradients. An obvious example is when a hungry animal wants to find food. If it detects some food cues, it will move in the direction of those cues. But now it has a dilemma: it wants to go toward the food, but it doesn't want to go on its own—if it leaves the rest of the group behind, it will isolate itself and expose itself to danger. In some animals, such as the mammals we'll meet in later chapters, there are clear hierarchies, with leaders who can dictate the direction of group movement. But in fish shoals and many bird flocks, there is no hierarchy and no permanent leader—the animals must reach a consensus. You'd think that would be almost impossible in a large group, but in fact it works well. Only a small number of hungry fish need detect food cues in order to move a large shoal toward the food. If something like 5 percent of the group starts to move toward it, the rest will likely follow.

Most members of the shoal have no concept that they are being led. So long as the majority have no strong preference for going in a different direction, they simply continue to align themselves with their near neighbors and respond to what everyone else is doing. The end result, of course, is that they fall in line. I once described to a group of students on a field trip how fish shoals and other animal groups could be led by a small number of motivated individuals. Ingeniously, a couple of the students decided to test this out for themselves, using the other students as their test subjects. Each morning of the field trip, these students would make their way from the field center in Pearl Beach, New South Wales, down a quiet country road to the seashore where they were working, and each evening they would retrace their steps. About halfway along this route, the road splits into two for about 150 feet, to go around a small island of trees and bushes. Without telling anyone what they were up to, the two students made sure they were at the front of the group as they reached the island, then decided at random whether to go left or right. Even though the group of some thirty students was strung out over perhaps 300 feet behind them, they followed whichever direction the sneaky experimenters chose. It didn't matter whether these leaders went to the left or to the right, the students behind them always followed. So it continued for several days, until the experimenters came clean and told the rest of the group what they had been doing.

If there's one thing that people hate, it's the idea that they're acting like sheep, so the revelation that this was precisely how they'd been acting made the group extremely uncomfortable. The next day, after they'd revealed what they had been doing, the two self-appointed leaders again went to the front of the group. They went to the left of the island, and everyone else went to the right, in a defiant show of independence. Except that, of course, the two leaders had still determined the route that the rest had taken. For if the other students wanted to show that the leaders had no influence, they should have walked to the left or right of the island at random.

People don't want to accept that much of what they do is governed by simple, often subconscious, responses to the behavior of others. Terms like *groupthink, herd mentality,* and *sheeple* have very negative connotations, yet, in many situations, subconscious social rules of interaction can be highly beneficial. Every time we use a busy pedestrian crossing, for example, we form into lanes, following the people ahead of us who are crossing the road in the same direction as we are. We aren't necessarily aware of this, and there doesn't seem to be any strict rule at work, such as "move to the right if about to crash into someone walking in the opposite direction." We simply self-organize. In the absence of any agreed rule, we respond to the social forces at work and form the most efficient solution to the problem. If we didn't, the outcome would be chaotic, with lots of collisions, and those awkward situations where you end up in an embarrassing pas de deux with a stranger, mirroring each other as you go right, then left, then right again, smiling ruefully as you finally break the impasse.

My PhD supervisor, Jens Krause, tried out a large-scale experiment on people that was based on our work with fish. He managed to get a couple hundred volunteers in Cologne one Sunday morning to participate in an experiment in a large hall that he booked for the purpose. Each person was given two simple rules that were based on those I described earlier in our experiments of fish collective behavior. These rules were "keep moving" and "stay within an arm's length of at least one other person." The second rule was of course a version of what the fish used as they moved, keeping close to a near neighbor.

What happened surprised even Jens. For the first few moments, people just moved about with no overall pattern, but quickly a kind of order set in as people found themselves moving in a large ring-shaped structure. As they realized this, some of the participants started to laugh, yet there was little anyone could do to break the pattern. No one had consciously tried to make it happen; the pattern simply emerged. More formally, this ring-shaped structure is known as a torus, and it is a feature of many kinds of animal collective movement, including schools of barracuda.

Next, Jens tried a variation of the experiment with a new batch of volunteers. The majority of participants were given the same two simple instructions as before, but a few randomly selected people were secretly given an additional mission—to reach a predetermined point at the edge of the hall. The question was, Could these few manage to lead the many, while still obeying the rule to remain within an arm's length of at least one other person? Sure enough, so long as at least 5 percent of the group was given the instruction to move toward the target, they were capable of leading the rest of the people who were in the dark about this objective.

It's probably important to point out that there's no magical, unwritten rule that says animal groups will follow if 5 percent of their number start to move in a particular direction. In small groups, you need a larger proportion. And this value will vary from species to species as well as for different situations. It simply demonstrates the point that you don't need a majority of group members, or anywhere near it, to agree before a move can happen. This means that animal groups can make decisions pretty efficiently. Even so, if 5 percent is roughly how many it takes to move a group, this still means that in a large group—say, of one thousand animals—you might need fifty leaders. It would be more efficient, arguably, if fewer leaders were needed to get the ball rolling. But although this might seem like a good idea, the fact that you need a certain number of animals to agree to the move provides insurance against the spread of bad information and the making of bad decisions.

GETTING IT RIGHT

Making decisions is an essential part of living in groups, and the evidence suggests that groups are usually extremely good at it, making accurate choices and reaching a consensus among their members. The more you think about it, the more incredible this seems. I've always found it amazing that animal groups do it so well, and finding out *how* they do it has been a major part of my career. After I finished my PhD with Jens, I went to work with Paul Hart at the University of Leicester. Perhaps it's no coincidence that at this time I was agonizing over some important decisions in my own life, trying to plan the future against the serial uncertainty of short-term academic contracts. I decided to look at how animal groups cope with this challenge, and I turned—with dizzying originality—to fish again. Well, I would, wouldn't I? Not only do I understand them, but they make excellent experimental subjects. You can test groups of them easily in the lab in a way that you would struggle with for any other social vertebrate. And you can generalize from them to other animals—the behavioral similarities are far greater than the differences.

Early in my time at Leicester, Paul introduced me to a favored collecting site, which bears the picturesque name of Melton Brook. A section of it flows fairly close to Leicester's city center. Perhaps it was also once a picturesque place, but it isn't now. Together with Mike, my companion from the later trips to the Lincolnshire ditches, who had joined Paul's group shortly after me, we christened it the River of Shit. The name barely did it justice—it was full of litter and smelled like the toilets at a festival, and we had to remove drink cans, wrappers, and discarded nappies to look for fish. Once, the bloated corpse of a dead rat came bobbing jauntily past. The River of Shit seemed to hold little prospect of supporting a population of fish, but in fact it was full of sticklebacks, which were perfect for my experiments. It also felt like I was rescuing them from an urban hell when I took them back to the aquariums at the university.

Some stories of animals making bad decisions have really caught the public's imagination. The most famous of these is herds of lemmings jumping off a cliff. Behaving like a lemming has become a byword for mindlessly copying a mistake. But lemmings don't do this. In the film that gave rise to the notion that they did, Disney's *White Wilderness*, Inuit children in the Canadian Arctic captured the lemmings for the filmmakers, who transported the animals to Alberta, where they were herded into a river for the edification of the moviegoing public. Thus, a myth was born.

Nonetheless, I wanted to know if the pressure of social conformity could make fish behave . . . like lemmings. Using my refugee sticklebacks from the River of Shit, I set up an experiment where small groups of fish would travel from one side of an aquarium to an area of cover at the far end. By dividing the aquarium for part of its length, I gave them two alternative routes, a bit like the island in the road where the students ran their secret human experiment some years later. To spice things up for the fish, I put a fake predator along one of the routes. Then I stood back to see how they handled their choice. Barring a couple of foolhardy fish, all of them avoided the predator route and took the safe option. No surprises there. Next, I tried to see if a bit of social pressure could influence them. I put in a replica stickleback and made it swim along the route past the predator. The real sticklebacks could watch their replica counterpart swim along this route before making their own choice of which way to go. The result was almost identical—they disregarded the replica and nearly all of them made the safe choice again. But I wasn't done yet. I wanted to keep ramping up the social pressure, to see what it would take to influence them. I now showed them two replica sticklebacks choosing the risky route, and suddenly the real sticklebacks took notice. Although they were far from unanimous in their willingness to follow these two pathfinders past the fake predator, there was a substantial shift in their behavior, especially if they were in smaller groups themselves. One "leader" was ineffectual, but two could change the way they all behaved.

In any group of animals, including humans, you can always get a reckless individual or a maverick that makes bad choices. For this reason, it's better to ignore what any single individual does—a simple rule like this acts as a filter for bad information. But two individuals doing the same thing at the same time? They might just be worth listening to. But that doesn't mean you have to follow inflexibly and slavishly, like the lemmings supposedly did. The fish in my experiments with two replica leaders usually set off as if to follow the two leaders, but plenty of them got cold feet (should I say fins?) and changed their minds to prefer the safer route.

We call this the *quorum response*: the tendency to ignore initiatives by single individuals, or, in some cases, smaller groups. It's a simple and beautiful means of avoiding mistakes. Grouping animals often wait before responding to a new piece of information until a quorum—a threshold number of group members—is reached; only then do they follow. In my experiment, I managed to get groups of sticklebacks to make a bad decision, but only by hacking their defenses, using the pressure of social conformity against them. In the real world, it's very unlikely that two real leaders would both have made the mistake of swimming past the predator in the first place.

In 1785, nine years before his career as a thinker was abruptly curtailed by the loss of his head during the French Revolution, Marie-Jean-Antoine-Nicolas de Caritat, Marquis de Condorcet (let's call him Condorcet from now on, or we'll be here all night) came up with a theorem. Condorcet was a mathematician and philosopher as well as, unfortunately for him, an aristocrat. Part of his legacy was a mathematical demonstration that large juries reached better decisions than small ones. This sounds fairly obvious, so, to do Condorcet justice, I should say that the problem is a little more complex than you might think. The rule, in more detail, is that if each juror independently arrives at an opinion on the guilt or innocence of the accused, then increasing the size of the jury typically increases the likelihood that when the jurors vote, their majority decision will be correct. In other words, you get better overall decisions as the jury size increases. Having twelve "good and true" people on a jury, as the courts of many countries do, serves this purpose well.

Choosing between two options is something that animals are often compelled to do. Getting it right is critically important to them. Early in my exploration of this subject, I heard what seemed like a weird statistic about US presidential elections. Ultimately, these boil down to a choice of two candidates. The statistic was that across every US election, regardless of policy differences and razzmatazz, the voters are significantly more likely to pick the taller of the two candidates. A quick fact-check showed this not to be quite true, although there is certainly a strong trend. Still, it got me thinking—about fish, naturally. If you gave fish a choice between two leaders, which one would they pick? With a bit of photoshopping, I gave groups of my urban sticklebacks choices of large versus small, fat versus thin, spotty versus unblemished, and so on. I did this by producing an image of each of these and showing them to the fish, and then moving the two images in opposite directions so that the fish had to make a choice of which leader to follow. Ridiculous though it sounds, each of these options has implications for the fish. Larger fish tend to be older and have more experience than smaller fish, fat fish may be better foragers than their thin brethren, unspotted fish tend not to be troubled by parasites, and so on. In each case, the slight differences between two potential leaders gave their would-be followers something to mull over.

What I found was that when offered a choice of leaders, the fish in small groups didn't show a particularly strong preference one way or another. But as the groups got bigger, their members were more and more decisive and made better choices—they usually reached a consensus quickly and united behind a favored leader. I didn't think much about this at the time; it was just a bit of fun and scientific curiosity, and I didn't know about Condorcet's jury theorem yet. It took a mathematician colleague, David Sumpter, to point out both my ignorance and the fact that this was a really nice demonstration, using fish, of the principle of how juries work. Who knew?

As the number of fish in a shoal increases, so their ability to make decisions improves. They become more accurate in their choices, and they make their decisions more quickly. This is a major benefit to fish, but it's not the only advantage of shoaling. Most fish lack spines or armor, and they're not poisonous; on the contrary, they're often delicious. In the absence of any form of physical defense, they have to rely on rapid movement to get out of the way of their hunters. Among vertebrates, fish devote the highest proportion of their bodies to muscle—it forms almost 80 percent of their overall weight, which is roughly twice the proportion of muscle in humans. This allows them to be excellent movers. But there's only so much that can be achieved by being quick and agile, especially in open water with few, if any, hiding places. Shoaling dilutes an individual's risk and allows fish to exploit the confusion effect, the dazzling sensory overload that predators experience. In addition, fish are better at finding food in groups, because the shoal acts as a kind of super-sensor, combining the searching abilities of the members of the group. When they're hungry, the fish in some shoals form a phalanx pattern that is wider than it is long, a little like a cordon of police advancing in a line as they search for evidence at a crime scene. When one fish finds food, the others home in on it to try and get a share.

SWIMMING INTO TROUBLE

The many advantages of shoaling help to explain why it is so wide-spread as a strategy. Of more than twenty thousand different species of fish, well over half gather into shoals for protection when they are vulnerable juveniles, while around a quarter of all fish shoal through-out their lives. In these fish, the urge to shoal is deep-seated: when they see a large group of their own kind, it lights up the preoptic part of their brains, which regulates social behavior. Incidentally, though there are many important differences between the brains of fish and those of mammals, including us, underlying it all are some funda-mental common elements—and in mammals, the preoptic area is important in social and sexual behavior. Among the fish that shoal throughout their lives are those most familiar to us on our dinner plates, such as cod, sardines, mackerel, tuna, and anchovies. The scale of the groups that some of these form can be almost beyond belief. A single group of anchovies in the Black Sea, for example, can occupy nearly 250 million cubic feet of water, and there are shoals of herring comprising hundreds of millions of individuals spread over more than 10 square miles. But rather like the krill, these vast aggregations are at great risk when they confront a new and very different kind of predator—humans.

For almost the entirety of their 530 million years of evolutionary history, fish have adapted to cope with the challenges posed by predators. But they have no answer to modern fisheries; or to modern vessels with hydraulic winches, which can deploy seine nets that are more than a mile long and 600 feet deep; or to trawlers that pull nets through the seas like giant mouths that are large enough to engulf a warehouse. Humans use sonar to take much of the guesswork out of the equation, homing in on large shoals. Gear like this not only overcomes the shoaling defenses of the fish, but actually uses it against them, exploiting their tendency to gather in numbers as a means of catching them all. Quotas may be levied to protect the health of the stock, but are not always either realistic or enforced. Bycatch—that is, the fish that are in the wrong place at the wrong time, and not even the species the fishing company is looking for—may simply be discarded. Ice machines enable the boats to preserve large catches on board rather than having to return to port with them, meaning that the boats can remain at sea and fish for longer stretches of time. Put simply, our ability to catch fish has increased beyond the ability of fish stocks to cope.

When the first Europeans to visit the Americas returned home, they described the astonishing riches to be had. The Venetian explorer Giovanni Caboto described the seas as being so thick with fish that they could be caught simply by lowering a basket into the water. Portuguese and Basque fishers, who were among the first Europeans to realize the potential of the Grand Banks off the coast of Newfoundland, have been crossing the Atlantic to fish there since the 1400s. British fishers visiting the area in the 1600s spoke of such an abundance of fish that they could scarcely row a boat through them.

The biological richness off the coast of Newfoundland is due to some unique oceanographic conditions. The Gulf Stream forges its way north from the Gulf of Mexico carrying warm water and collides off Newfoundland with a southward flow, the cold and nutrient-rich Labrador Current. At the point where they collide, the currents are pushed upward as the seabed rises to the shallow waters of the Grand Banks, an area of the sea roughly the size of Ireland. As they rise and mix, the currents draw more nutrients with them, supporting a rich community of microscopic life that forms the foundations of a productive food web.

At the apex of this food web is Atlantic cod. This fish is emblematic of our troubled relationship with marine stocks. We prize them for their succulent white flesh, and each fish can have a lot of flesh: cod can grow to the size of an adult human—up to 6 feet long and weighing more than 200 pounds, though monster fish of these sizes are seldom seen these days. Female cod are capable of producing mind-boggling numbers of eggs. The largest females in a population may be twenty years old and produce some ten million eggs per year, whereas their smaller, younger sisters might produce less than a tenth of this number. If you take the largest fish from the population, the ability of that population to bounce back suffers disproportionately, yet inevitably, the largest fish represents the greatest prize.

Aside from the egg-producing potential of the larger cod, the experienced fish take up leading positions in migrating shoals. In the early 1990s, Elisabeth DeBlois and George Rose, two researchers from Canada's Department of Fisheries and Oceans, followed a single, massive shoal of cod as they performed their annual migration to the north of Newfoundland. The shoal stretched for over 10 miles, with the largest fish in the vanguard. The younger followers gain from this arrangement by learning long-established routes, so knowledge passes down from generation to generation. A similar pattern has been seen in the mega-shoals of other food fish, such as herring, mackerel, and sardines. Taking the largest fish therefore has major drawbacks for the population as a whole.

The Atlantic cod bounty on the Grand Banks seemed endless, and for centuries, it effectively was. Even as the British colonized Newfoundland and fishing communities sprung up around the coast, there was little impact on cod numbers. There were so many cod, and the fishing capabilities of humans were so limited that the fish population didn't suffer. However, the balance began to tilt in the early 1900s. Fishing vessels no longer relied on wind power, but on engines, first coal-powered steam and then diesel. Wooden boats gave way to larger craft with metal hulls. Manpower at the winches gave way to horsepower from the engines. A greater number of vessels were drawn to the Grand Banks, joining the small-scale local fishers. Catches slowly but inexorably increased.

Then, in the mid-twentieth century, things went haywire. The boats and the nets they could haul had grown, but that wasn't the only problem: the development after World War II of onboard refrigeration heralded the birth of the super-trawlers. The Canadian government could claim exclusive rights to the waters within 12 miles of their coast, but beyond that it was a free-for-all. Factory trawlers descended on the Grand Banks from all over the world to share in the spoils. Eyewitness accounts from Newfoundlanders at the time describe a blaze of lights from the fishing boats out to sea, as though a city had been founded there. The catch peaked in 1968, when almost 900,000 tons of cod were taken from the Grand Banks; in the years following, the cod fishery declined sharply even though the fishing intensity remained the same. By 1974, the catch was less than half of what it had been in 1968. The cod population was dwindling fast.

In 1977, the Canadians increased their territorial limit to 200 miles, but there was little respite for the cod, as the locals now sought to profit from the bonanza. The logic was simple: Why shouldn't Canadian fishers claim what was theirs, when foreigners had been reaping the benefits for years? Even though by now the catch was a fraction of what it had been at the peak, it was a boom time for the fishers. Quotas were set, but they were wildly optimistic about what the fishery could sustain. Warnings from small-scale inshore fishers, who noted that both their catches and the size of the fish they were taking were down, went unheeded. The political will to stop the fishing juggernaut was lacking.

By 1990, the situation was critical. George Rose, a scientist on board a government research vessel, described how he and his colleagues located a huge group of cod, almost 500,000 tons of fish traveling south from the Arctic to the Grand Banks. It was likely the last cod mega-shoal—Rose estimated that perhaps 80 percent of the remaining stock was concentrated in that one group. Two things can happen when the population of a shoaling fish like the cod is in steep decline. Either the number of shoals stays roughly the same, but each shoal gets smaller, or the number of fish in each shoal stays the same, but the number of shoals decreases. George believed the latter was closest to the truth. And so he could do nothing but watch as that last shoal, which had migrated down from Arctic waters, reached the fishing grounds and swam straight into the welcoming embrace of a phalanx of trawlers.

The end came soon afterward. In 1992, the Canadian government finally bowed to the inevitable and announced a moratorium on cod fishing on the Grand Banks. It was estimated that 99 percent of the breeding stock of cod had been removed from the area. At a stroke, thousands of people lost their jobs, and the cost to Canada through lost revenues could be measured in billions of dollars. The more immediate costs to people's lives in Newfoundland is incalculable. It's a stark example of the tragedy of the commons—the cod bounty was at once owned by everyone and no one. There was no incentive for the fishers to curb their take, because whatever they left behind would be scooped up by someone else. Any political determination to take decisive action in those critical years in the run-up to the collapse simply wilted in the face of the risk to a successful industry. Besides, cod don't vote.

When it was first announced, the moratorium was intended to last only a couple of years, allowing the stocks time to recover. But there was no immediate bounce-back in cod numbers. It looked as though the ecosystem had been altered irrevocably, not only by the effective removal of the key predator, but also by the damage that the trawlers did over time to the seabed, the nursery both for cod and for their prey species. As years passed with still no sign of recovery, there was no alternative but to extend the moratorium. Then, in the mid-2000s, the first signs appeared that the cod were coming back. The numbers were small, but they were there. Even now, almost thirty years since the original moratorium, the population is at most a third of what it was in the boom years. Intense pressure is nonetheless being exerted by the fishing industry, and quotas have been increased as a result. Too much, say the scientists. Too little, say the fishers. No one knows what the future holds or whether we are capable of learning from our mistakes. The only certainty is that the Grand Banks cod collapse was a tragedy for humans and fish alike.

Shoaling is integral to the success of many species of fish, yet it was the downfall for the northern cod of the Grand Banks. You might ask why the cod didn't abandon their tendency to shoal. Given enough time, or, more specifically, enough generations, it is not impossible that cod behavior would have evolved and adapted. But their impulse to aggregate and their strategy of migrating in groups are fundamental parts of their behavior, shaped over tens of millions of years. When factory trawlers came along with their vast nets, the cod faced an onslaught so new, so rapid, and so devastating that they were unable to cope.

Despite the impossible situation that faced the cod, animals can change their behavior and adapt it to their needs. Fish may shoal more strongly at some times and be more independent at others. For those species that are active during the day, the onset of night is the trigger for their shoals to break up. The reason for this is shoaling's role in confounding predators, and particularly predators that hunt by sight. In the dark of night, these predators are no longer so active, and the protection afforded by the shoals is no longer needed. At first light, when danger threatens once more, the prey fish coalesce back into their groups. The greater the danger, the tighter and more cohesive the shoals. Observations of minnow shoals in an English river suggest that these fish can live their entire lives within 10 feet of their predators. Even so, minnows are among the most successful fish in the habitats where they are found, which shows just how effective shoaling is as a defense. If the danger is ever lifted and the fish can relax, their need to shoal decreases.

TRADING PLACES

The streams that flow through the steamy rainforests of Trinidad provide one of the most fascinating examples of evolution at work in the natural world. This Caribbean island is home to the guppy, a familiar resident of many a fish tank. These are small fish, no more than an inch or so in length, with a relentless, unquenchable commitment to sex. This isn't for kicks, though; it's a survival strategy for a species that lives a short and perilous life. Their rapid breeding made guppies an instant hit in the early days of the aquarium trade and facilitated the efforts of those who, by means of artificial selection, felt they could improve on nature. The gaudy results of linebreeding can be seen in pet shops: male guppies with long, colorful, flowing tails, a trait shared to a lesser extent by the females.

But in the wild, guppies are far from the fluff-finned monstrosities that infest aquariums. They're much more subtly patterned, and with good reason: in Trinidad, they live side by side with a battery of different predators. A domestic guppy would be a sitting target for any predator worth its name—their colors make them stand out, and because of their exaggerated tails, they can only swim slowly compared to their wild cousins. While diligent breeders have selected the most colorful individuals in each batch of young, and used these to propagate subsequent generations, predators select in the opposite direction. Instead of nurturing the most glamorous, they eat them. You can see the results of this difference in an unintended natural experiment in Australia, where over the years foolish fishkeepers have released their fish into the waterways of the tropical north. Guppies are highly adaptable creatures, and they've taken to their new habitat with an enthusiasm that now makes them a problematic invasive species. It's reasonable to assume that the guppies that were originally released were the domesticated and highly colorful variety, yet their descendants look almost exactly like their Trinidadian, wild-type counterparts.

Native Aussie predators have shaped this transformation over successive generations. Meanwhile, in Trinidad, as streams make their way from the mountains and hills to the sea, they flow over waterfalls. Above the falls, guppies live a life of comparative ease, free from large predators such as the pike cichlid and the blue acara, which are confined to the waters below. Those guppies that are swept over the waterfalls have a much tougher life ahead of them, a world of risk and danger. The guppy population is thus divided by a natural barrier into two distinct sets of living conditions: on the one side, a kind of fishy heaven, and on the other, a predator-infested hell. The guppies, happily, are adaptable enough to cope. A combination of individual flexibility and natural selection has resulted in clear differences between these two populations. Those above the falls are dandies—larger, more colorful, and less inclined to hang around in groups; those below are comparative urchins, drab in appearance, and they prefer to hide from their persecutors in shoals. The downstream males can't be too drab, though—females prefer brightly colored mates, so to have a chance of breeding, the males must express vivid shades of orange and black, even though this makes them stand out to the predators. As with so much in the natural world, it's a trade-off. The guppies below the falls live fast and die young; surrounded by ravening predators, they breed at an earlier age than the ones above the falls, cramming their entire existence into a few short weeks.

This natural crucible of evolution has attracted biologists for decades. The different pressures on these coexisting guppy populations have over time produced a diverse range of traits. What happens if you switch fish between the different habitats? Animals tend to be flexible in their behavior, adapting to different challenges throughout their lives. Just like us, however, they adapt better to change as youngsters. If low-predation guppies are raised alongside high-predation guppies, they tend to behave like them, and, in particular, to shoal more—in other words, they conform to the behavior of the animals around them.

More lasting changes, such as aspects of behavior that are under the control of genes, take longer. If you breed guppies from different populations in captivity, you can raise them in controlled conditions. This allows you to separate "nature" from "nurture." Nature describes those parts of an animal's behavior that are controlled by genetics, while nurture refers to how an animal's environment affects its behavior. In the studies, guppies who were separated from their wild populations by two generations still behaved something like their grandparents, despite living in safe and controlled conditions. In other words, captive guppies whose grandparents were from high-predation sites still shoaled more than their counterparts whose grandparents were from low-predation sites. Their behavior is, in part, genetically predetermined. So although there's some flexibility, a guppy—or any animal—isn't a blank slate, free to change its behavior to fit its surroundings. There's carryover from previous generations. By carrying out a type of analysis that geneticists call a *heritability estimate*, we can look at just how much of an animal's behavior comes down to its genes. This has been done for shoaling behavior in fish, and the answer seems to be about 40 percent. That is, across a fish population, about 40 percent of behavior is determined by genes, and the remaining 60 percent by the environment, or at least by nongenetic factors.

We don't often get to see evolution at work in complex animals like vertebrates in the wild; it simply takes too long. But the speed with which guppies zoom through their life cycle means that the process in these little fish happens in fast-forward. Back in 1957, some guppies were taken from a high-predation environment to a new stream that lacked predators. A similar transplant was done in 1976. When the descendants of these guppies were examined again in 1992, their behavior had changed. The behavioral characteristics that each subsequent generation inherited from its parents had been modified by natural selection. As a result, we know that it takes between thirty and sixty generations of guppies to change their behavior from that typical of guppies in a high-predation environment to that of guppies in a low-predation environment. What we're talking about here, of course, is the removal of a predator threat, so the selection isn't as intense as it would be the other way around. Guppies introduced into a low-predation environment will have plenty of time to breed, whether they lose their hypercautious nature or not. If fish from the low-predation sites were introduced to high-predation environments, the selection pressure would be more extreme. Because only those individuals who were able to adopt effective anti-predator behavior would survive long enough to breed, the rate of evolution would be greater.

If you or I were to see a tiger on Main Street, the chances are that we would invest a little effort in trying to put some space between ourselves and it. Not so for guppies, or for a great many other shoaling fish. (Of course, it's not tigers they're seeing, but their fishy equivalents, yet the point is the same.) Rather than hightailing it to safety, they keep a watchful eye on their nemeses. But why would they do this? For one thing, the predator you can't see is the one that poses the most danger. For another, information is a vital currency for all animals.

Even so, what happens when prey fish first detect a predator seems incredible. They first stop what they're doing to focus on the threat, and then, very often, a small group, or even a loner, may leave the shoal to approach it. What they're doing is called *predator inspection*. It's a risky business, so it must be done cautiously. As the spies close in on the predator, they pause frequently to assess the situation, and to allow other members of the inspection party, if there are any, to keep up with them. They are careful to avoid the so-called attack cone of the predator—the area right in front of its toothy mouth. They know that the hunter must turn before it can lunge, which buys them a little security.

The purpose of the mission is to get an idea of the level of threat the predator poses. As they inspect it, they gather information, such as whether the predator is hungry, and what kind of prey it has recently eaten. That information can be gleaned from various cues from the predator, such as whether it looks full, and how it smells. A hungry predator with a taste for your own kind is bad news, so it's time to be wary; one that's stuffed to the gills is less of a problem. Armed with the fruit of these observations, the inspectors report back to their shoal. If the news is good, they can get on with their normal activities, such as browsing and courting.

Since approaching a deadly threat is inherently dangerous, not every fish in a stream is willing to do it. So why do some members of the shoal put themselves in harm's way, and run the gauntlet of inspection? For lots of animals, risk-taking can be sexy. Studies aimed at teasing apart the complexities of human sexual behavior consistently point to risky behavior as being attractive. This is most evident in humans in assessments of male attractiveness by women, but only when the risks relate to our hunter-gatherer origins: for example, confronting dangerous animals or putting out a fire is sexy, but more modern risks, such as not wearing a seat belt, are not. A similar phenomenon happens with male guppies returning from predator-inspection visits. Their reward for taking on this risky business is increased attractiveness to the females and greater receptivity to their courtship attempts. Normally, female guppies base much of their mate-choice decisions on the males' color patterns. Males that are strong and healthy and that are good foragers can flaunt brilliant colors. Their livery is a signal to the females that they have good genes. What chance is there, then, for drab males in this world? Engaging in risky predator-inspection behavior turns female heads to such an extent that they prefer risk-taking but drab males over timid but brightly colored ones.

A TRIP TO PARADISE

Twenty-five years on from my boyish wonder at Aysgarth Falls, I'm on a new adventure. I'm on board the *Heron Islander*, a 100-foot catamaran, departing the port of Gladstone in Queensland to go to the

Great Barrier Reef for the first time. I'm more than just on board, I'm right at the prow, as if being at the very front of the boat will get me there quicker. I gaze out over the 50-mile stretch of the Coral Sea that lies between me and the most famous marine habitat on Earth, willing the boat on. I feel like the luckiest person alive. This is the culmination of everything I've worked for. As a fish researcher, it's my Shangri-La.

As the catamaran eats up the distance, the sea changes color from deep blue to bright turquoise as we pass Masthead, then Wistari Reefs, the landward outposts of the Great Barrier Reef. Finally, we arrive at Heron Island. This was once the site of a factory that caught and canned turtles. They don't do that in Australia anymore—they put beans in cans and leave turtles to their meanderings, which I think is the right way around. Heron is now a resort island, but it's not the end of the journey. Waiting at the jetty, there is another boat to take us the last 12 miles to our destination, One Tree Island.

Russ and Jen, the managers of One Tree Island Research Station, are skippering the second boat. It's plain sailing until you reach the outer fringes of One Tree Reef. This must be crossed before we can reach the lagoon, but it can only be done at high water and at one specific place. Even then, there's little room between the hull of the boat and the precious, yet boat-destroying coral. Getting through requires a mix of nerve and nautical skill, both of which Russ and Jen have in spades. Today the sea is calm and we cross into the lagoon easily. Just a mile or two more and we're at the island. One Tree is a coral cay, essentially a pile of millions upon millions of coral fragments. Over time it has been colonized by hardy vegetation and small trees. I don't know how many trees, but it's more than one. One Tree is pretty small, though—only about 4 hectares (10 acres) at high tide. Most of this is off-limits to us—we're in one of the most pristine areas of the reef and conservation is at the top of the agenda. Access to the station is limited to researchers, and since the island lies in one of the most stringently protected zones of the reef, all fishing is prohibited. The research station has expanded from a simple hut in the early 1970s to a small collection of buildings including an office, a lab, and sleeping quarters, but even so, it is huddled in one corner of the island to leave the rest free for the wildlife.

Even though One Tree is remote, plenty of animals have washed ashore and established themselves over the years. Geckos, spiders, and enormous venomous centipedes have been there as long as anyone can remember. Birds, of course, face no difficulty in island hopping. Rails run through the undergrowth, and egrets wait by the water's edge to spear unwary fish. Thousands of terns gather there and make their untidy nests in the crook of just about every branch of every tree. Here on One Tree, they have never learned to fear humans; they just stare at you with silver-fringed eyes that look for all the world like they're wearing spangly makeup. They don't flee as you pass by, but they might peck you on the nose if you get too close.

They have learned to fear the birds of prey, though. A pair of white-bellied sea eagles nest on One Tree, and when one of these takes to the wing, a huge alarm goes up from the terns. At dusk, a small number of muttonbirds arrive. The eerie mewling of these feathered visitors in the darkness used to terrify superstitious sailors, who thought they were being haunted by the ghosts of dead babies.

As wonderful as the birds are, I'm here for the underwater animals. No sooner have I dropped my bags than I'm striding out into the water. There are something like 1,500 species of fishes on the Great Barrier Reef, and I'm keen to make their acquaintance. I dive below the surface, and as the bubbles clear, the first marine animal I see is a huge loggerhead turtle. It turns out that I've met Astro, a large male named for the covering of turf-like algae on his carapace. Astro regards me dolefully, then goes back to feasting on a giant clam, biting through more than an inch of tough shell as though it were a wafer. Venturing on, I see more kinds of fish in an hour than I'd seen in the rest of my life. It is everything I've dreamed of and far more.

The activity is dazzling, offering limitless possibilities for my research. Everywhere I look there is something going on—hiding, hunting, disputes, pursuits, display, foreplay, and foul play, all happening within my sight and repeated onward out through the lagoon. I could take my pick, but since I'm most interested in social behavior, there is one species that particularly draws me. Dotted across the sandy floor of the lagoon are clumps of coral ranging in size from a flowerpot to a garden shed. Some stand on their own in glorious isolation, while others are clustered together into undersea gardens. Each of these corals is a mixed mass of living polyps and the limestone skeleton that provides them with their support, while also offering a labyrinth of hiding places for fish. As I pass each of the smaller clusters, a collection of little faces peek out at me from the safe recesses of their home. Intrigued, I put a little distance between myself and their hidey-hole and wait. After a minute or so, the owners of these faces emerge and buzz around, rarely straying far from their sanctuary. These are groups of damselfish, and their intense black and white stripes, which resemble an old-fashioned mint candy, give them their name: humbugs. Each of the coral clumps has its own group of humbugs; typically only around six of them cohabit in each. They are small fish—the largest in the group would fit comfortably on the palm of my hand. Usually, each group contains a range of sizes, from the boss right down to the tiniest, glorious in its miniature detail and sometimes no larger than your fingernail.

Humbugs are interesting because they're so different from most other social fish. The first and most obvious difference, which I've already alluded to, is that normally groups of fish are matched in size, but humbugs are not. This leads to the second unusual aspect of these fish, which relates to their behavior: they have a very strict dominance hierarchy. Every fish knows its place, and if it forgets, the others soon remind it. The largest fish is usually a male, and his smaller housemates are female—but like many coral reef fishes, humbugs are hermaphrodites. They begin life as females and change to become males when they reach full size. In the group, the aggression of the male prevents females from developing into males, and so he keeps a tight control of his harem. This is the exact reverse of the situation of the humbugs' close relative the clownfish, made famous as Nemo in the Disney film, where the female is the largest and keeps the smaller males in check.

Another difference between humbugs and most other social fish is that, for most of the time, the humbug group is a closed shop. They don't leave the group; nor do they welcome newcomers. This means that the group is very stable, staying together at their home coral for months or even years. If a vagrant comes along, the established fish in the group that most closely matches it in size will try to drive it off. No humbug wants to lose its place in the queue to an outsider, but the larger ones won't be displaced by a new group member and the smaller ones aren't big enough to fight it. So it falls to the fish most directly affected—the one closest in size—to defend its rights.

Like the majority of coral reef fish youngsters, humbug larvae disperse to the open sea once they hatch before returning a few weeks later to find a home. These tiny fish face enormous odds against their survival in this time, but if they manage to locate a humbug colony, they are usually allowed to join as the most junior (in every sense) member of the group. Larger females may depart the group to escape the tyranny of the male. Then they will transform into males themselves and seek a group of females to join them, and so become tyrants themselves.

With all this strife and aggression in their groups, you might imagine that humbugs wouldn't work well together as a collective. Coral reefs, for all their beauty, are one of the toughest habitats to live in. Humbugs and other small fish are faced with a near-constant threat from an army of different predators. Taking your eye off the ball even for a moment can be fatal. I once watched as a pair of humbugs came spiraling out of the coral, locked in a furious dispute. A predatory wrasse saw its opportunity and gulped them both down. Their combat had made them blind to danger for just a moment, which proved to be the last mistake they'd ever make. That was the exception, though. Normally in the face of danger, humbugs work together against the common enemy. Every member of the group has its role to play in this effort, and with so many eyes on the lookout, it's hard for a predator to sneak up within striking distance. As soon as one humbug spots approaching danger, it reacts with a sudden dash to safety in the coral. This dash alerts the rest of the group, regardless of whether they have seen the predator, to its approach. In the blink of an eye, all are safe within their refuge.

In this age of social media, there's a worrying trend toward what are known as *silos*. This is where groups of people come together online, reinforcing their own biases; soon their opinions converge. This behavior forges a rewarding group identity, but has the negative consequence of making them impervious to alternative viewpoints.

We can perhaps see the foundations of this tendency in the humbugs. Their small social circles are inter-reliant and long-lasting. Little wonder, then, that the group shapes how its members behave. Although each group behaves differently from others in its locale, within any particular group each individual fish acts very much like its groupmates—it responds to challenges and to opportunities in a characteristic, socially set way. Humbugs join established colonies at a young age and then most often live their entire lives within the same group. Their behavior patterns are molded by the other group members as they grow, with the result that their individuality gives way to uniformity. The fish in each group aren't related, so their behavior is the result of their social environment; it's nurture rather than nature. Although we can see obvious downsides to silos in our society, silo-type social behavior melds the humbugs together as a group, so that all are on the same page: their consistency and conformity become valuable weapons against predators.

In the wild, humbug groups might have a territory that includes two or three neighboring corals. The safest way to move between these is together, to the point that the group seems to be an extension of the humbug individuals. I wanted to take a closer look at how these little fish coordinated their behavior to make decisions as a group, so I collected a few of the groups and took them to the aquarium rooms at One Tree. First, I gave each humbug colony a fairly basic bit of coral, a kind of starter home. Then, once they'd settled in,

I tempted them with a highly alluring humbug mansion at the far end of the aquarium. Less social animals than humbugs would probably go over for a look one by one, but not these fish. Although they were interested in inspecting the new real estate, they would only approach it as a group, so, to start with, they remained anchored at their starter home. I watched as the decision process unfolded, and noticed how the activity—you could almost call it excitement—of the group built as they swirled around their primitive base. Those that were impatient to view their potential new home started swimming more excitedly, sometimes in circles, sometimes feinting as if to cross to the new house before retreating if no one went with them. Finally, when all of them were elevated to fever pitch, they'd cross as a group.

On a couple of occasions, a group would cross, but leave one of their number behind. The isolated individual grew agitated on these occasions, but not for long—spotting that one of their number was missing, the others would return home, as a group, and scoop the stray back into their ranks, and then they would cross together to the new luxury residence. This buildup to a group movement is by no means a humbug-only phenomenon—we see it in lots of group-living animals, from flocks of geese waiting to take to the air to horses and even gorillas. The rise in activity preps all the group members, which is critical for ensuring group cohesiveness; when the time comes to embark on the journey, all the members of the group are ready. Simple though it seems, it's a precursor to the social behaviors manifested by birds and mammals.

HUNTING AS A SHOAL

Not all fish groups are formed for the protection of their members. Sometimes fish come together to hunt, and, when they do, the results can be amazing. Goatfish typically live in small, stable groups, and their hunts show the hallmarks of cooperation. One or two members of the group will chase after the quarry, a smaller fish, and, on the reef, their prey will most likely disappear into a coral head. When this happens, the other members of the group move to the sides to block its escape until the prey is surrounded. The goatfish then probe the coral with the whiskery barbels below their mouths. If they can spook their target, it may dart out, straight into the mouth of someone on the welcoming committee. Unlike other cooperative hunters, the goatfish don't share their prey. But it's a good strategy so long as cooperation provides greater shares to all than hunting on their own would yield.

Few fish have achieved the notoriety of the red-bellied piranha. These South American freshwater fish grow to the size and shape of a dinner plate and have large, razor-sharp teeth. Like their close relatives, the small and inoffensive tetras that inhabit many an aquarium, piranhas live in shoals. Their reputation as frenzied killers of man and beast is founded on a memoir written by former US president Theodore Roosevelt about his travels to the Amazon in 1913. Keen to impress the visiting dignitary, the locals took an unusual and gory approach. They herded a cow into the river and, as Roosevelt watched, the waters boiled as piranhas stripped the unfortunate bovid to the bone in a matter of minutes. Roosevelt's account of the events of that day was relayed around the world, and the piranha took its place as one of the most feared animals on the planet.

Is its reputation deserved? Well, no, not really. The grisly fate of the cow was staged. To ensure an unforgettable experience, the local people had netted off a stretch of river and filled it with piranhas. The piranhas were separated from any food source—although they're not averse to eating one another—for an extended period before they pulled their stunt. Starved of food and given a weakened animal to awaken their appetites, it's perhaps not surprising that the result was a bloodbath.

We shouldn't conclude that piranhas are entirely harmless. Their fearsome dentition and powerful bite can certainly do some serious damage, and there have been tragic, recent cases of their bites killing people, as well as many more instances of them causing injuries, especially to hands and feet. But feeding frenzies are not really typical, and taking on prey larger than themselves is unusual, unless their

quarry is already weakened. Normally, they hunt other fish, sometimes even taking bites out of some of the huge catfish with whom
they share their home. The tendency of piranhas to live in shoals
seems to have far more to do with protection, just as is the case with
so many other social fish species. Piranhas make a tasty meal for cai-
man, cormorants, and other, larger fish. In small groups, they show
signs of anxiety in common with other prey fish that feel vulnerable,
especially rapid breathing. The refuge provided by being in larger
groups has a calming effect on them, so these terrifying hunters of
legend are actually banding together for safety.

By contrast, sailfish aggregate only to hunt. From January to
March, vast shoals of sardines move north to feed on plankton
blooms at the Isla Mujeres bank off the Yucatán Peninsula of Mexico.
The arrival of the sardines brings a host of predators, including the
sailfish. These are phenomenal marine hunters, reaching up to 10 feet
in length. The first quarter of their bodies tapers to a point, sometimes called a sword or a bill, but more properly known as a rostrum.
They use this in an extraordinary way. To find out more, a team of
biologists, led by Jens Krause and including Alex Wilson, chartered a
boat from Cancún.

How do you find sailfish in open sea? The answer is that you look
for the birds. Wherever sailfish are feeding, the commotion attracts
feathery fish hunters, such as frigate birds, who soar above and then
plunge into the shoals to collect their share of the bounty—the sailfish's prey. Aggregations of these birds act as markers that can be seen
for miles. Thirty miles out to sea, Jens's team found what they'd been
looking for.

In the water, the aggregations of sardines had gone deep to avoid
the perils they faced in the brightly lit surface waters. Out of sight of
the researchers, the sailfish had pursued their quarry to the depths.
Alex described the swishing sounds of their swords as they harried
the sardines far below and tried to corral them to the surface. Their
goal was to separate a clutch of sardines from the main shoal. If they
succeeded, the sardines were in real trouble. In situations like this,

adrift from the millions of their own kind, the little group of sardines comes under attack from all sides. There is nowhere to run and nowhere to hide—except among their own—so they stick close to one another, turning fruitlessly this way and that in their panic. A group of sailfish will concentrate on an isolated baitball of sardines. Their pursuit is relentless, but it is orderly. They take it in turns to approach. Their frantic prey dart away from them, but they're most often no match for the speed of the sailfish. Catching up to the rear of the sardine shoal, a sailfish will put its rostrum in among them, then slashes to one side. The rostrum is studded with tooth-like ridges, and as it slashes to the side, it inevitably wounds a sardine, knocking off scales and gashing its flesh. Momentarily destabilized, the injured sardine has milliseconds to recover and rejoin the shoal before it gets wolfed down by its attacker. But there is no respite. Another sailfish will repeat the trick, and another sardine will be injured. As the wounds grow and exhaustion sets in, the sardines become easier to attack. Gradually, the shoal of sardines is winnowed down. When only a few are left, the orderly pursuit of the sailfish breaks down and there's a free-for-all.

There's little refuge in the open ocean for the sardines, but they can be resourceful. Sometimes the relentless attentions of the predators cause the sardines to abandon the ever-contracting baitball en masse to form a living blanket around one of their persecutors. An inquisitive shark, drawn by the commotion of the hunt, finds itself coated in terrified sardines. If they avoid the shark's mouth, this is a safe enough place to be, for the time being—the sailfish won't risk breaking their rostrums against another large predator.

Occasionally, the sardines cluster against the watching researchers for the same reason. It's mostly a short-term measure; they're merely delaying the inevitable. That said, Jens and Alex did come across a sardine that survived against the odds. Spotting something floating in the water, Alex swam over to investigate. It looked like a dead turtle, but there was a single sailfish circling it. As Alex approached, the sailfish glided away. But the dead turtle turned out not to be so dead after all. It withdrew its head from its shell and shot off, leaving a single sardine, which had been hiding under it as a last resort. In a desperate bid for survival, this sardine had kept the sailfish at bay. With the turtle gone, it dashed downward, back toward its fellows in the shoal far below. It had lived to tell the tale.

CHAPTER 4

CLUSTERFLOCKS

Birds of a feather not only stick together,
they go on crime sprees.

MURMURINGS

Some years ago, before I realized that I could earn a crust studying animals in the name of science, I had an uninspiring job in the northern English city of Bradford. Each day's drudgery was bookended by a plod to and from the train station. But one late November evening, that commonplace commute was transformed into something incredible. I can still see it in my mind's eye now. As dusk falls in Bradford, I leave the office to head to the station, huddled, like everyone else, in coat and scarf. There's a gloominess to late autumn in any English city. The nights draw in, the pavement and buildings are perma-damp, there's the bite of coming winter in the air. Usually, the only animals are a smattering of tough dogs and gnarled pigeons, hoping for someone with cold hands to fumble a kebab. But as I reach Forster Square, a great cacophony of noise from above draws the attention of everyone around. Right over our heads, a vast flock of starlings performs an awe-inspiring aerial ballet.

It's a mesmerizing, shape-shifting spectacle. Each bird seems to be in tune with some mysterious choreography, so that the flock wheels, dives, and climbs as one, pulsing with kinetic energy. Their calls are so loud, they drown out the traffic. For a few minutes, we earthbound creatures are privileged to see one of the most wonderful mass displays in nature, a starling murmuration. Finally, the starlings are ready to roost. This final act of the evening's show is initiated by a handful of leaders beginning their descent. With each bird taking its cue from its near neighbors, the move propagates rapidly. In near-perfect synchrony, the murmuration leaves the skies.

The show that the birds put on in the skies over Bradford that evening is played out in many different places during the coldest months of the year, October to March. In some places the starling gatherings are enormous, containing up to a million birds. Accordingly, the Danes refer to the aggregations as the *sort sol*—the black sun. The birds congregate to perform their coordinated maneuvers for around half an hour as dusk begins to fall before descending en masse for the night. It's thought that flocks like this recruit from afar and coordinate the actions of their fellow starlings. And perhaps, by huddling together after landing, they can stave off the cold weather.

But the main reason for the formation of the massed flock is predators. A single starling, or even a small group, is vulnerable to raptors, such as sparrowhawks and harriers, but there is strength in numbers. The murmurations seem to be larger and longer lasting in the presence of these mortal enemies. Such a dazzling display of whirling, speeding multitudes may serve to confuse the hunters, stymying their attempt to select a victim.

Predators are confounded when the flock moves in coordination, but in such huge aggregations—of hundreds or thousands—how can each bird monitor the behavior of all its flockmates and keep up? The answer is that they can't. In fact, we now know that each bird is responsive to only around seven of its nearest neighbors. Paying close attention to those nearby is something that all animals do when they coordinate their behavior. We do it ourselves when we walk and when we drive. For the starlings, it's an essential measure to avoid a catastrophic midair collision, and it allows each bird to align its behavior with its close companions. Why seven? Like so many things in life, that number marks a trade-off. On the one hand, the more neighbors a bird pays attention to, the better its ability to coordinate its behavior with them and to respond to sudden turns; on the other hand, the higher the number of birds to pay attention to, the more difficult it is to keep track. It turns out that responding to six or seven near neighbors makes for a perfect balance between these two factors. It allows the birds to maintain maximum responsiveness at minimum cost.

Despite each starling only paying close attention to the birds nearest to them, the flock as a whole can change speed and direction almost instantaneously. It's a staggering feat, and to explain it, we need to delve into the idea of criticality. This term describes the point where a system is on the edge of making a transition from one state to another. Snow accumulates, becoming ever less stable until the apparently calm and picturesque mountainside abruptly transitions into a deadly avalanche. Invisible tectonic plates push against each other, building up energy until a critical point is reached, the plates shift, and an earthquake results, as if out of nowhere.

This same idea, borrowed from physics, might just explain the behavior of starling murmurations. The flock, on red alert for an attack, is constantly near criticality, poised on the edge of making a sudden transformation in its flight path. If one bird makes an abrupt change in its flight, the whole flock transitions. So each bird influences all the others as information, in the form of changes in flight direction and speed, zooms across the flock. Another remarkable thing is that it doesn't seem to matter how many birds there are; the ability of the flock to coordinate remains, which is vitally important when it comes to confounding predators' attentions. Those of us watching the starling flock as it morphed and reeled on that cold evening in Bradford saw a wonderful example of animal artistry, but for the starlings it might have been a matter of life and death.

GOING LONG-DISTANCE

As they gather in groups to fly, birds form two familiar yet distinct kinds of flocks. One of these is what we might, with a wry smile, call clusterflocks—dense clouds of birds with no clear configuration, just a series of endlessly shifting patterns. Starling murmurations are one spectacular example of these, but there are many others, particularly among smaller species of birds, which often come under the threat of attack by birds of prey as they fly.

The other kind of flock flies in formations that starkly contrast with the freewheeling shapes of the clusterflocks. Larger birds, such as ducks, geese, and swans, form into regimented V-formations, especially on long-distance journeys. One bird must always be at the apex of the V, leading the way, a role that is rotated at regular intervals. The leading bird in the formation acts something like the riders at the front of the peloton in the Tour de France. It takes the brunt of the wind resistance while providing important aerodynamic benefits to those following behind. This is why the birds (and the cyclists) have to switch positions frequently to take the strain of leadership. Fitting birds with heart-rate monitors, similar to the technology that many people now wear on their wrists, shows just how important these aerodynamics are. Birds in trailing positions have heart rates around 10 percent lower than the leader. It may not sound like much, but flying over long distances can take birds right to the edge of their capabilities. Something like a third of young snow geese die during migration. So for a bird, saving even a fraction of your energy might just make the difference between arriving safe and sound or running out of fuel on the journey.

Why do the geese and ducks fly in a V-formation? As birds flap their wings, they push air downward. This downwash causes other pockets of air to rise, and the upwash is greatest immediately behind the downwash and slightly off to each side. So if birds fly behind and to the side of their mates, they can ride the rising air and thus save energy. This much, at least, we knew from theory and the study of single birds flying in wind tunnels. But research still needed to be done to work out the details, and the holy grail until recently was to see if groups of free-flying birds conformed to our laboratory-formed expectations. All-around bird genius Steve Portugal, from Royal Holloway, University of London, took on that challenge in his quest to get to grips with bird migration. His dilemma was that, although it's straightforward enough to fit birds with tiny data loggers to collect detailed information as they fly, retrieving them is no easy task once the birds have flown for hundreds or thousands of miles.

The solution came from conservationists in Austria, who were trying to reintroduce the marvelous—though, let's not mince words, tragically ugly—bald ibis to its central European range centuries after it went extinct there. Reintroducing a migratory bird is no simple task. For one thing, birds learn their migration routes from their elders in the same population, so when there are no elders to teach them, you have to figure out how to show them the way. The Austrians were doing this using microlight aircraft, which the ibis were trained to follow. Steve decided to also study the ibis—because it, too, flies in formation—and to use the same method to guide them along their route; once they'd landed, he could collect the data.

Now that they had the technology and the study species, Steve and his team could get to work and learn just how V-formations work in free-flying, migrating birds. What they found was nothing short of astonishing. Although they expected the birds to benefit from the small pockets of upwash created by the birds in front of them, they'd assumed that their ability to hit exactly the right spot would be a bit haphazard. This hypothesis proved to be an insult to the birds. Not only did the ibis conform to the predictions of where they would position themselves relative to the birds in front of them, but they moved their wings just about perfectly to keep riding the upwash patterns in the air. The wing of one bird traced almost exactly the same path that the bird ahead of it had traced a moment before. Steve compared it to how children walking in deep snow follow the foot-steps created by the adults walking ahead of them. Except this is in the air, with wind and turbulence to account for and no visible path to follow—the precision needed to do it is extraordinary.

Another amazing element of the ibis's flight was how they appear to cooperate with one another when it comes to taking on the tiring lead role in the V. In a bicycle race like the Tour de France, cyclists' roles are clearly defined: some bear the brunt of the wind for much of the race, while others ride in their shadow to save their energy. Theoretically, ibises could do something similar, bullying subordi-nates into doing more than their fair share at the front of the group. What they actually do is to share the workload in an admirably fair manner, with each bird spending the same amount of time traveling in the energy-sapping front positions as well as coasting in their part-ners' wake. No bird is given—or takes—a free ride.

Our understanding of bird migrations has itself been a long jour-ney, appropriately enough. Strange as it may now seem, the appear-ance and disappearance of birds at certain times of year was once shrouded in mystery. Where did they go? Natural philosophers bent their minds to the task and came up with a range of entertaining suggestions. Aristotle asserted that they went into hibernation, quite possibly underground, or changed magically into different species. Others scoffed at these notions and proposed that they disappeared underwater. One admirable free thinker suggested that they flew to the moon.

It wasn't until two centuries ago that we managed to consign these excellent theories to the rubbish bin. Across parts of Europe, large, twiggy constructions appear on chimneys and rooftops every spring. These are the nests of white storks, a bird with a 6-foot wingspan and a saber-like bill. Each year, these large birds travel to their European breeding grounds and begin weaving their rather chaotic abodes. In 1822, a white stork showed up in Germany with a spear through its throat—one that was clearly identifiable as being African in or-igin. The hunter's aim had been true, and yet the plucky bird had still managed to fly all the way to Germany, crossing the Sahara and the Mediterranean with its dramatic, bespoke piercing. It must have thought that the worst was over, right up until it was shot on arrival in northern Europe. The bird was stuffed and now takes pride of place in the zoological collection at the University of Rostock, a monument to terrible luck. Twenty-five more such "arrow storks" were subsequently reported, which finally solved once and for all the question of where the storks overwintered.

Even without the encumbrance of a projectile through the throat, traveling long distances is a challenge for such hefty animals as white storks. To help themselves, they make extensive use of thermals to gain height. Thermals are caused by the sun heating up the ground. Some patches of ground heat up more than others, which causes a column of hot air to form there, and it rises high into the atmosphere. When they find such a thermal, white storks circle within it, gaining altitude. They then glide along to the next thermal, where they can regain their height before another glide. The birds effectively use these air columns as stepping-stones, or fuel stops, on their migration.

The problem is that thermals can be tricky to find—they're not visible to the naked eye, and while cloud formations help, it can still be hit-and-miss. Like the storks, airline pilots are also on the lookout for thermals, but unlike the birds, aviators try to avoid them, since they're a cause of the dreaded turbulence that might mean you end up wearing your in-flight coffee. Yet even with an array of instruments at hand, pilots often rely on information passed on from earlier pilots following the same route to know where problematic air patterns are likely to occur. As it turns out, flying in groups allows the storks to collect information in a similar way. Once a bird finds one of these giant midair elevators, it will start to loop around in the column of air, a signal for other birds to join it.

But the discovery of a thermal isn't the end of the story. Shifts in the wind cause these columns to dance around like smoke from a bonfire on a blustery day. Collectively, the storks have the answer to this, too, tailoring their own positions according to those of the group and to their reading of the capricious rising air. The net result is that all use the conditions effectively, rising like a living helix of birds, and gaining altitude for the next stage of their long journey.

DO MANY WRONGS MAKE A RIGHT?

It's fair to say that the West of England Fat Stock and Poultry Exhibition of 1906 is better remembered in the annals of science than many other, no doubt equally prestigious, fat stock exhibitions of the time. Present at that Edwardian country fair in Plymouth was an elderly scientist and statistician named Francis Galton, a man who lived by a motto: "Whenever you can, count." Galton's attention was drawn to a competition to guess the weight of an ox after it had been slaughtered and dressed. The entry fee of sixpence deterred casual guessers, for whom 6d would buy around three pints of beer at the time. Nonetheless, something like eight hundred people entered the competition, filling in a card with their guesses, names, and addresses. Of those, thirteen entries had to be discarded because their cards were unreadable—presumably some of those entrants had already had their three pints of beer. Galton managed to get hold of the remaining cards and made some calculations. It turned out that the correct answer for the weight of the ox was 1,198 pounds, and that the average of all the guesses was 1,197 pounds—together, the Plymouth crowd had guessed the weight of the ox with impressive accuracy.

At this point, it would be entirely reasonable to ask what on Earth guessing the weight of an ox has to do with birds. What it demonstrates is the ability of a group, with the information it has collectively, to achieve astonishing precision. It doesn't necessarily matter if some of the participants are wildly inaccurate—if there are enough guesses, then groups are often remarkably good at making these kinds of predictions. In particular, the group as a whole typically does better than any one individual. This phenomenon, known as the *wisdom of the crowd*, or *swarm intelligence*, is the basis for the effectiveness of search engines, such as Google. But it is not restricted to human groups. Animal groups can also benefit, and bird flocks provide a good example of it.

When birds depart for their migration, they use one of a host of different cues to set their route, including the Earth's magnetic field, the position of the sun or stars, smells, low-frequency sounds, or major landmarks such as mountain ranges and coastlines. Even with this capability, each individual bird might be slightly inaccurate in reckoning the course. If this is so much as a degree or two out, over a long migration it could mean missing the landmass they are heading for. If they migrate together, however, they can pool together all the information they have at their disposal in a way that has some parallel with how the Plymouth crowd guessed the weight of the ox. Although each bird has a slightly different preference for the direction of flight, so long as the flock keeps together, in theory the birds should tend to travel in roughly the average direction of all the birds' directional preferences. The result of this collective navigational trick is that the flock as a whole ends up in the right place.

That's the theory, but does this happen in practice? The evidence suggests that it does. For example, flocks of skylarks and scoters, a kind of sea duck, seem to navigate more effectively when they fly in larger groups than when they fly in smaller ones. Perhaps the best evidence comes from pigeons. These birds are so commonplace and so little regarded nowadays that it's easy to forget their seemingly miraculous ability to find their way home. It wasn't always like this. Pigeons are thought to have been the first type of bird to be domesticated, and they have been used as messengers for centuries, carrying vital communications, medicines, and even contraband material with unerring accuracy at a flight speed of over 60 miles an hour. Although they've largely been supplanted by the telephone, and, more recently, by the Internet, they're still used in some remote areas of the planet. Indeed, they were of sufficient concern to the Taliban in Afghanistan and to ISIS in the Middle East that people were banned from keeping them. Pigeons still have their devotees, and a diverse list it is, too: noted pigeon fanciers throughout history include Elvis

Presley, Mike Tyson, and Queen Elizabeth II. While the practice of using pigeons as messengers has largely faded into history, the sport of pigeon racing is still going strong, to the point where birds change hands for tens of thousands of dollars—one went recently for $1.9 million.

Although pigeons are clearly capable of navigating on their own, they are social birds and perform better in groups. Because of recent breakthroughs in GPS technology, we can study the routes traveled by migrating animals and so start to understand this incredible behavior for the first time. To examine whether pigeons gain any benefit from navigating in a flock, versus finding their own way, researchers fitted some birds with miniature GPS devices and tracked them as they faithfully returned home from a release point. The results were clear—when pigeons flew in groups, they traveled by the fastest, most direct route back to their roosts. Not only that, but they traveled faster in groups than they did on their own. Pigeons that were released on their own made it to their destination, too, but first they circled for a while, apparently getting their bearings. Then they navigated home in a different way, following landmarks, even though this method didn't necessarily take them on the shortest route.

Researchers in animal movement refer to the *many-wrongs principle*, which means that navigational errors on the part of individuals are corrected in a cohesive group via collective wisdom. It's a neat solution to the problem of working out where you should be going. But it's perhaps not surprising to learn that birds don't solely rely on it, especially when mistakes can mean the difference between life and death. A problem with the many-wrongs approach is that you could end up with a group of birds that all have the same incorrect bias in their preferred direction for traveling together. We can make an analogy here with peoples' political beliefs—if you only listen to those with the same opinions as yourself, all that happens is that your bias is strengthened. For migrating birds, it's vital that every member collects and updates its own information, correcting its course as it does so. Across the whole flock, the birds integrate and respond to direction changes, and group navigation maintains its effectiveness. Not only that, but flocks seem to be able to function as giant airborne sensors. If one of their number detects a faint cue and changes its speed or direction in response, the rest of the flock will usually follow suit. In this way, you have a large number of animals each searching for clues to help guide them to their destination. When one of them finds such a clue, all of them benefit from the find. It's a simple and brilliant way of navigating.

HITTING THE BOTTLE

Having links to other individuals allows the spread of ideas. In my office at the University of Sydney, I sit with my back to a window overlooking a busy thoroughfare, flanked by trees that are in turn thoroughfares for birds. Two years ago, during an unseasonably damp spring, a bedraggled and dismal-looking bird turned up on my windowsill and regarded me mournfully. I consider my office to be relatively well equipped, but it was clear at that moment that I lacked bird food. For want of anything else, I tried the one thing I had on hand: porridge oats. The bird gobbled them down. That was the start of a friendship which, if not beautiful, might at least be described as functional, and it is still going strong. "Ken" is a noisy miner, named for his unfashionable dominant color, gray. They are an unpopular species in Australia for their tendency to attack people during the nesting season. As an unpopular species in Australia myself, a Pommie, I can identify with that, although I don't attack people, breeding season or no. The thing is, noisy miners are sociable birds, and once word got out that food was to be had at Ward's Kitchen, I found myself playing host to more and more of them. I was forced to keep my window shut. Ken is the only one who knows the password, which is to tap gently on the glass, and the backup password if I fail to respond, which is to scream at me. But with the close links within noisy miner social networks, I guess it's only a matter of time until that secret spreads, too.

Perhaps the most famous example of natural learning in the animal kingdom was a rather grander affair than a handful of birds visiting an office. It comes from a lovely, if unheralded, bird known as a tit. Around a hundred years ago, reports came in of birds from the tit family prizing open the wax tops of milk bottles left by early-morning milkmen on the doorsteps of houses near Southampton on the south coast of England. By doing this, the birds were able to get a free lunch, drinking the cream off the top. Necessity is, as always, the mother of all inventions: the birds were most likely to do this in winter, when food resources were scarce and the rich cream was of most value to them. Within just a few years, this piece of avian ingenuity spread throughout the British Isles, passed from bird to bird by a process known as *social learning*. One bird watched another solving the problem of how to get at the cream and, at some level, thought "what a good idea," before imitating the behavior on a different doorstep.

Keen to find out just how widespread this behavior was, the British Trust for Ornithology sent out questionnaires to its members around the country, to branches of natural history societies, and even to the press, asking whether they'd seen similar behavior and, if so, when they'd first noticed it happening. Because the trust kept up its interest—and the questionnaires—throughout the first half of the twentieth century, we can see how the behavior spread. Tits are small birds, and they tend to be homebodies, staying fairly close to where they grew up—yet the outbreaks of milk thievery were separated by hundreds of miles. In places as far apart as Coventry and Llanelli, a small number of isolated cases soon mushroomed into a widespread daily dairy heist. Outbreaks of thievery spread rapidly through the local tit populations, following the paths of main roads between suburbs as birds went door-to-door for a cheeky meal.

In some parts of the country, local law-abiding birds never lowered themselves into the criminal classes. In Little Aston, near Birmingham, the practice was common, but it failed to spread to the nearby towns of Streetly or Sutton Coldfield. In other parts, the birds were

absolutely shameless, following the milk cart and attacking the bottles before they were even delivered. Attempts to dissuade the larce-
nous birds by placing stones or inverted cans on top of the bottles
rarely held them up for long. Having solved the initial problem of
how to get past the bottle caps, these keen little brains soon overcame
the new obstacles as well.

We can't be sure whether the birds learned from one another or
each bird solved the problem on its own. And if they learned from
one another, we can't be sure whether it all started with one brilliant
bird, and traveled from village to village, or multiple birds had the
same idea and passed it on just within their neighborhood, causing
the copycat behavior locally. And we can't be certain that the spread
of the behavior was due to the information being shared between
members of the tit population, or simply to people being more vigilant once they heard about it, and then noticing the behavior, or to
more questionnaires being filled in over time. Although the spread of
tit criminality makes for an excellent anecdote, it doesn't quite pass
muster as evidence of social learning. Even so, the way the patterns
spread in different parts of the country, starting with a single outbreak and radiating outward, first slowly and then gathering pace,
does at least suggest that the great avian milk heist was based on
social learning.

A more grisly example of tit ingenuity is provided by the brain-
eating birds of Hungary. Although tits are not equipped with the
fearsome weaponry of some others of their kind, such as birds of
prey, they recognize an opportunity when they see it. The caves that
hibernating pipistrelles use also serve as a winter larder for the great
tits. As the bats begin to awaken from their long sleep, their calls attract the attention of the tits, who home in on the sounds. Still dozy
from hibernation, the bats provide an easy meal for the birds, who
peck through their thin skulls to get at the succulent brains within.
As with the milk-bottle tits (whose misdemeanors now seem so much
less sinister), there is a suggestion that this behavior is transmitted
between birds and down through the generations.

The ability to watch and learn from the actions of others allows animals to access the accumulated wisdom of their population. The milk-drinking and bat-biting exploits of tits do suggest social learning, but how might we be sure? What would it take to demonstrate it in a scientifically acceptable way? The answer is by conducting experiments that systematically exclude alternative explanations for a given pattern.

A couple of years ago, a team led by Lucy Aplin from the University of Oxford set out to do just this. Using a number of different populations of tits as their study system, the researchers provided the birds with a problem. A tasty mealworm was on offer, but only if they could solve the puzzle of how to break into a feeder to access it. To help them on their way, some of the birds were caught and given an intensive course on how to open the feeder. The feeders had two doors, a red one and a blue one. Either way worked, but some birds were taught that they could open the feeder by sliding the blue door from left to right, while others were taught to slide the red door in the opposite direction. A third group of birds were shown the feeder but not how to open it. Then they were released with their newfound knowledge back into the wild to mix with their local population, while feeders were installed in anticipation of what was to come.

The results showed beyond any doubt not only the birds' capacity for social learning but also how the information spread. Even though only two birds with the knowledge of how to open the feeders were released into each population, three weeks later around three-quarters of each population had learned the trick. The birds that had been shown the feeder, but not how to open it, were also released into their populations, and they didn't fare quite so well. Still, tits are pretty decent problem-solvers, and even the populations that hadn't been chosen for box-opening lessons gradually worked out how to get their beaks on a mealworm—though nothing like as effectively as the populations seeded with graduates from the feeder-opening school.

Other interesting findings emerged from the experiment. Even though there were two ways of opening the feeders, sliding the blue door or the red door, the population adopted and stuck with whichever solution they had been taught, even though both methods were equally rewarding. When, through trial-and-error, the tits worked out that there were, in fact, two ways of opening the boxes, they still preferred to stick with the approach they had learned from their fellows in the same population. The birds are traditionalists, just as susceptible to social conformity as we are.

After four weeks of experiments, the feeders were packed away, to return the following winter. It's a sad fact that small passerine birds like tits don't tend to live long. Some nine months after the feeders had last been seen by the birds, only a little more than one in three of the problem-solvers were still alive. Despite this, the remaining birds hadn't forgotten. They took to the feeders with relish, and the process of social learning from experienced to naive began again. Remarkably, the preference for how to solve the puzzle had strengthened in this time—the tradition of being either blue-door birds or red-door birds was even more firmly entrenched.

TOOLED UP

The beautiful Pacific islands of New Caledonia lie around 800 miles off the northeast coast of Australia. To Captain James Cook, the glorious white, sun-soaked, palm-fringed beaches immediately called Scotland to mind, and thus the islands got their name. Perhaps I lack his imagination, but I couldn't see it myself when visiting in 2018. Still, I put the disappointment to one side, grabbed a facemask, and consoled myself with a spot of snorkeling in the brilliant, clear waters. A couple of hours later, I had sated, at least temporarily, my appetite for fish-spotting. Drying myself, I went looking for the animal for which the islands are most famous, at least to nerdy biologists— the New Caledonian crow. It's not much to look at, admittedly: just a medium-sized black bird. But just to prove that brilliance can reside in the most unassuming places, it does something that only a handful of animals can—it uses tools.

I followed a rough track into thick forest, feeling almost immediately the splendid isolation of the explorer. My eyes darted this way and that as I kept a sharp lookout for exotic beasts of any description. In such an abundance of fecund nature, I was sure to see something wonderful. A few minutes later, deep in the forest, the track took me to a clearing where I found a disgruntled cow tethered to a stick, surrounded by rubbish. No doubt Captain Cook would have conjured a fitting name, but as the cow and I looked at one another we seemed to share a sense of disappointment, and nominative inspiration escaped me. Daunted, I carried on. At one point, I might have heard the "qua qua" call that the crows make, and which gives them their local name, but ultimately I failed to see the famous birds, or their tools. Fortunately, others have.

New Caledonian crows rewrote the rule book when it comes to animal intelligence. They are the only non-mammal that not only uses tools but passes on their innovations and modifications from generation to generation in a phenomenon known as *cumulative cultural evolution*. The crows feed on juicy grubs, but these are hard to get at, hidden deep within crevices and holes in vegetation. Taking a long, tough leaf typical of a tree such as the screwpine, they determinedly snip away at it with their beaks until they have crafted an avian work of art. The ideal utensil has a combination of length and grip. The length extends the crows' reach and enables them to access grubs from out-of-the-way places. The naturally serrated edge that remains when the leaf is stripped away gives them purchase on their prey, but if that's not enough to winkle out the reluctant insect, they'll make a hook. It's a deft and flexible approach to the kind of problem-solving that for so long has been thought of as the exclusive preserve of humanity.

Examination of the tools left by crows across the islands gives us an idea how these amazing birds gradually develop more efficient grub-defying paraphernalia over time as they learn from one another.

Because there is some variability in the tools they use in different parts of New Caledonia, it seems as though engineering knowledge may be passed among birds in a local area, and particularly within family groups. New Caledonian crows aren't the most social of the corvids, but a pair of adults might accompany their offspring for a year or so as the youngsters grow—long enough to pass on their tool-making heritage. They might just have something to teach us as well. In most animals, humans included, the knowledge tends to flow one way, from old to young, from teacher to pupil. There can even be resistance to learning from juniors—you can't teach an old dog new tricks, especially if you're a young dog. But when small groups of New Caledonian crows have to solve a new problem, the older birds are just as likely to learn from the younger ones as the other way around.

Social learning isn't only about foraging strategies or finding your way. The astonishing variety of calls and songs made by birds are shaped by the community in which they live. Birds use their vocalizations to advertise territorial claims, attract mates, and spread the alarm. Exactly how they do so can vary from place to place. In our own societies, one of the most obvious influences of a social environment can be seen in the adoption of local dialects and accents, which act like badges of group identity and membership, and are partly related to the desire to emulate others and fit in. These local inflections develop relatively early in life. Children are quite flexible in this regard; in immigrant families, they can develop accents that sound noticeably different from those of their parents. At a young age they can also adopt the speech patterns of foreign languages far more intuitively and effectively than adults. My own son, Sam, made friends with a Chinese boy when he was three years old. Although he'd never spoken Mandarin before, his ability to imitate the boy's language, and especially his pronunciation, surprised the Chinese boy's parents with its accuracy. But this is a narrow window, and once it passes, it is nearly impossible to sound convincing to a native speaker in a second language. By the time we reach our teens, our flexibility in this regard is all but lost. Although there are always exceptions that prove the rule, by that age we are usually a finished product in terms of our spoken accent.

Birds are subject to similar kinds of social influences, so it's not surprising that they, too, have local differences in phraseology and pronunciation. Young white-crowned sparrows in the forests of North America acquire their repertoire of songs by listening to their neighbors during the first two or three months of their lives. The result of this localized social learning is the development of distinct dialects. Groups of these sparrows in different parts of the country and even in different forests have distinct calls. This doesn't just matter for fitting into the social scene—it makes a difference when it comes to sex—female sparrows generally prefer males who sound similar to themselves. The strength of this selection means you can even find birds of the same species that live pretty much side by side yet have different vocalizations. In parts of California, groups with different dialects may be separated by a few yards. In these cases, young males cunningly learn both dialects, thereby hedging their bets when it comes to impressing females.

COMING TOGETHER

Some of the most fascinating and dramatic examples of avian sociality are provided by gatherings of birds as they come together to breed or sleep. Often the individual birds trace a lonely path through their daily or seasonal lives, yet converge at times to mass at a single place.

At Bempton on the east coast of Yorkshire, towering cliffs rise sheer for some 300 feet. They are bulwarks against the North Sea, scoured by northeasterly winds and battered by thunderous waves. On the tops, the few, hardy shrubs reel back from the assault, careening like broken ships to landward. These headlands around Flamborough hold a special place in my heart, as it was here that I used to come as a child with my family for summer holidays, and here that as a town dweller I got a first glimpse of the breathtaking diversity of nature.

Yet walking on the tops of the cliffs in the middle of winter, these cold, salty walls of rock seem almost devoid of life. It's hard to imagine anything eking out an existence in such a place. For all that we know to expect change with the cycle of the year, the contrast just a few weeks later is scarcely believable. The cliffs are transformed into a raucous, exuberant carnival of activity as half a million seabirds return from afar to breed. The birds have spent months dispersed across huge areas, but as the urge to nest takes hold, they converge on Bempton. The cries of kittiwakes, guillemots, fulmars, razorbills, and puffins may not have the melody of songbirds, but their energy is magical. The majestic gannets are here, too, but these large birds remain slightly aloof from the rest, favoring the high-end real estate offered by the flat tops of limestone stacks that stand in defiant isolation after the softer rock around them has eroded and the cliffs have retreated. No matter how many times I visit this seabird city, watching the birds as they wheel about the cliffs or perch precariously on ledges, it never seems less than extraordinary that they succeed in nurturing eggs on these shallow and vertiginous ledges. The lives of their chicks, braced high on the cliff in niches of less than a hand's width, seem full of danger. Yet there's a paradoxical safety in this precarious habitat. Few predators can negotiate the cliffs. Although herring gulls do harass nesters, occasionally snatching eggs and chicks, the cliffs provide the breeding birds with both a refuge and access to the rich foraging grounds they need to feed their hungry broods.

Why do the birds gather here in such numbers? There's an element of tradition—it makes sense for birds to return to the place they first hatched, or where they've been successful in raising chicks in the past. A similar tradition draws me back there whenever I can go, even though I now live on the other side of the world. The pull is some vague, but strongly felt, sense that part of what made me will always be rooted in this place. Perhaps the birds experience a similar impulse, in their case one that's tempered and reinforced by the knowledge that it's a place that has proved itself—there's no need to change a winning recipe. Alongside this, for colonially breeding birds, a crowd draws a crowd. Even in those species that don't show much inclination to live with their own kind for the rest of the year, company exerts a potent attractive force. By homing in on the sounds and smells of a colony from far afield, or being guided in by other returners, inexperienced birds seem to gain reassurance from the presence of so many of their kind—as if this surely means that it's a good place to raise a family.

These cues are powerful. Just how powerful is suggested by a study of young bobolinks, a kind of small blackbird that is drawn, perhaps irresistibly, toward the audio playback of sounds from older, more experienced birds. So persuasive is the draw, in fact, that it overrides their own instincts about what makes a good breeding ground.

During the breeding season, all parental birds are under intense pressure, with ravenous and apparently insatiable chicks to feed. Perfecting the art of foraging and learning the best places to find food takes time, however, and often it's only the most savvy and experienced of parents who succeed in raising a brood. This might be one reason why many seabirds have an unusually prolonged developmental stage before breeding. Gannets, for example, get progressively better at foraging as they age, and wisely wait five years before starting a family. They don't have to rely solely on their own wits though—gannets are keenly aware of what their neighbors are up to. A gannet returning to the colony from a successful foraging trip attracts the attention of its peers, inciting them to head out to sea and retrace its path to a promising fishing ground. There's good reason for this—seabirds, like many colonially nesting birds, feed on animals that are concentrated in patches, such as shoals of fish. There might be miles of empty ocean hiding just a few dense shoals of sand eels or other prey fish. Even so, some areas of ocean are reliably better than others, and that's where experience counts. When a knowledgeable bird sets out to go fishing, it might attract followers, keen to piggyback on the veteran's know-how. The colony, then, acts as an information center for birds. By taking note of birds who have been successful, others can jump on the bandwagon. It can take some of the guesswork out of foraging.

Crammed into every available nook and cranny of the chalk cliffs at Bempton, the seabirds seem to completely saturate their habitat. Yet for all the seabirds might think they're congested, weavers can still teach them a thing or two about cramming a few more in. Weavers are a group of small birds related to finches. Their name comes from their incredible nest-building exploits. Males construct elaborate nests by—as their name suggests—weaving, knotting hundreds of strands of vegetation together to form intricate architectural masterpieces. The nest is usually spherical, rather than the cup-shaped structure made by most birds, so weavers get protection from above as well as below. Sometimes they also construct a tube, a bit like an entrance hall, that serves to restrict access to unwanted guests. The effort that goes into building these nests doesn't go unrewarded—females choose their breeding partners on the basis of their beautifully woven constructions. But when he finishes building his orb-shaped temple of love, the male weaver doesn't rest; he gets right on building the next one, making up to fifty in a season. If for whatever reason a nest doesn't impress any females, he will sometimes tear it apart and doggedly start over.

Many of the weavers are gregarious and build close together to form colonies. Of these, the sociable weaver of southern Africa builds massive nests that can be more than 350 cubic feet in size—that's about the size of a VW camper. Developments on this scale can hold hundreds of birds and are rather like avian apartment blocks, with separate chambers for each pair. These spectacular bird buildings can last for decades, housing generation after generation. Not only do they provide protection from predators, but they also buffer the extremes of temperature outside, a fact that isn't lost on other animals, including skinks, beetles, mice, and even other birds, who often move in to live alongside the weavers.

A bird apartment block is one thing, but the weavers' congregations are nothing compared to the roosts of the now extinct passenger pigeon. The largest of these could include billions of birds spread over more than 75 square miles. Historical accounts have described birds clustering in such numbers that every inch of space was taken up. Undaunted, the pigeons would simply sit on top of each other. Often there were so many birds that tree branches as thick as your leg would break under their weight. And underneath, pigeon dung would accumulate like stinky snow to the depth of a foot. Birds that flock often roost together, while birds that feed alone seldom do. Although passenger pigeon roosts represent an extreme, there are many birds that aggregate in their hundreds or thousands. When there are no nests to provision, or chicks to feed, birds that are sometimes quarrelsome in the breeding season are happy to sit alongside one another in the off-season.

Tree swallows spend much of the day on the wing, alone or in loose groups, chasing airborne insects. But each night, they seek out the company of their own kind to rest, often in impressive numbers. Flocks of these birds start to congregate around an hour before nightfall in marshland or stands of trees. Like the birds seeking a colony to nest in, swallows are attracted to the presence of their own kind, and the flock draws them in from miles distant. And since the same birds often return to the same roost each evening, there's a doubly strong pull to the area. The flock sweeps through the habitat until at some unknown signal they start to settle. The roost offers safety in numbers. Even though the commotion that occurs at a large roost may attract predators, the swallows pick roost sites that provide some measure of protection from ambush, and they check out the area carefully for signs of danger before settling each evening.

Coordination is an essential part of social living. It's part of our everyday lives and our calendars, from daily working patterns and mealtimes to annual festivals and holidays. For birds, roosts provide a rhythm that helps them to retain the cohesion of their social existence. They form each evening and break up each morning (or the other way around for nocturnal birds, such as long-eared owls). Like colonies, roosts act as information centers, giving birds vital intel about likely foraging sites. Tradition plays a role in dictating where they come to rest, too—they often use the same place to rest each night and are drawn to their own kind through social attraction. Some birds exert more of a pull than others. Just as in human society, where celebrities are paid to endorse products, birds are not immune to the attractions of status. Once the top-ranking birds decide on a place, the younger ones may be drawn to it. This is smart, because older, more experienced birds often have the best information on roosting and foraging sites. (Whether the same can be said for celebrity endorsements is another matter.)

For those high-status birds who attract a following, there may be a benefit to drawing a crowd. Often, the top birds can be found in the most desirable place in the roost, whether that be in the center of the congregation or toward the tops of trees. Those birds that cluster around them, or below them, act as a buffer against predators. Those below also pay the price of being crapped on. This is not only rather humiliating; it also compromises their feathers, making them less effective both for flight and for insulation. Cue some heavy-duty preening . . .

For quite a number of species, roosting occurs most often during the coldest months, which suggests that the birds also benefit by gathering together for warmth. This is about more than just comfort: birds can become sluggish in low temperatures, which can make them an easier target for predators. Again, occupying the prime spot is a valuable perk. As well as having a living shield against opportunist hunters, the birds in the middle of a group get the most heat insulation. Long-tailed tits line up along tree branches at night, with the head honchos in the center and the less-well-favored exposed to the elements at the ends of the line. In a cold winter, each bird may lose a tenth of its weight every night as it tries to keep warm, but the most exposed birds suffer disproportionately.

Social insulation like this approaches an art form in the high latitudes of the far south. It doesn't get much colder than in the penguin rookeries of Antarctica, where emperor penguins may experience temperatures of −40°C (−40°F). To survive, the birds have to maintain a core temperature of around 37°C (98.6°F), so there's an obvious need to insulate. This is partly achieved by the penguins' ultra-fine downy feathers beneath their outer layer. They also have a kind of heat-exchange system that keeps warmth from being lost in the blood flowing to the coldest regions of their bodies. Radiative cooling is at work as well—the outermost parts of a penguin's body are actually colder than the air around them, and so take up heat. Emperor penguins even avoid putting their feet flat on the ice much of the time, resting instead on their equivalent of heels.

But as valuable as these adaptations are, huddling together is essential. The icy winds can draw warmth from their bodies, but by huddling in tight groups, the emperor penguins gain some respite from the elements. In fact, so effective are these huddles that the birds in the center may start to overheat. As a result, penguins switch position frequently as hot and bothered birds in the center move to the edge and their chilled confederates move to the interior of the huddle.

BIRD SOCIETY

While many birds may congregate as they feed, fly, sleep, or breed, there are some that take this further, developing the strong relationships that form the basis of a true society. Lots of birds form lasting pair bonds, and though they may spend much of the year apart, they reunite and reaffirm their vows at the outset of each breeding season through a series of intricate rituals. But few can match the commitment to family life shown by Florida scrub jays. Scraping a living in the tough conditions of their natural habitat means that everyone has to muck in. In the case of these attractive blue-and-gray birds, it involves living in extended families. This is so familiar to us in our human societies that we may take it as the norm, but it's rare in the natural world. Young Florida scrub jays are raised not only by their parents but also by their older siblings, who, rather than fly the nest, remain with their families, helping out with the domestic chores. There can be up to eight adult birds at home in the territory—the two parents and six of their adult offspring, the so-called helpers-at-the-nest. These helpers play a vital role, collecting scarce food for the youngest members of the family and joining in with the never-ending task of defending their patch from predators and the next-door neighbors.

All of this seems a very civilized and recognizable state of affairs to us, yet to the helpers it may simply be a necessary compromise. Perhaps there's a parallel in our own societies, with the increase in young adults being forced to stay at their parents' homes by student debt and sky-high property prices. Helpers might stay at home because their options are limited—in the scrub jay housing market, too, it can be tough to get on the ladder. If they play the long game, the helpers may one day inherit the family residence. In the meantime, there's the consolation of supporting their younger siblings, even if it means delaying their own breeding chances. It's telling that across the range of species that exhibit this stay-at-home behavior, if an opportunity to set up on their own comes up, the helpers very often grab it.

Helpers are so valuable that parent birds sometimes demand their continued presence, however, so staying at home can be not so much a choice as a compulsion. The fathers of young African white-fronted bee-eaters act like evil characters in a Victorian melodrama, sabotaging their sons' attempts to leave home and breed independently. They do this in various ingenious and appalling ways, including obstructing junior's courtship attempts, or, if that fails, blocking the young couple's entry to their nest, and even turning up at their nest to bother them. Faced with this constant harassment, the younger males may simply give up and go back to the family business. And the helpers themselves sometimes resort to underhanded tactics. Young male bell miners help their fathers out with the task of feeding youngsters, but, just occasionally, their halo slips a little. They go through the motions of bringing food to the nest before rather sneakily eating it themselves. And it *is* sneaky: they tend to do it when no one—except the nonplussed chicks—is watching.

PECKING ORDERS

The term "pecking order," referring to a social dominance hierarchy, is so widely known that few people stop to consider its origins. We routinely use it to describe hierarchies in our own lives, especially in the workplace, even though we seldom peck one another. Consequently, people are often surprised that it comes from studies of chickens made almost a hundred years ago by a dedicated Norwegian with a name like something out of a Norse saga, Thorleif Schjelderup-Ebbe. Thorleif's boyhood fascination with the social dynamics of the occupants of his parents' chicken coop ultimately led to him studying them for his PhD.

He noticed that certain birds always got the prime spots for roosting and first access to food. He also noticed that they defended their privileges aggressively if other chickens tried to muscle in. By watching the patterns that these bouts of aggression took, he realized that there was a hierarchy among the chickens. The dominant chicken would peck at any bird that tried to assert itself, while the next in the hierarchy would avoid taking on the dominant bird but peck at those lower in the hierarchy—and so on, to the unfortunate bird on the very bottom rung, who had to settle for the most meager rations and poorest shelter. So the term "pecking order" originally was quite literal.

Although we might have mixed feelings about hierarchies in human society—usually depending on our place in them—in animals they can be beneficial. They have the perhaps surprising effect of reducing aggression within a group. Once established, chicken pecking orders, for instance, tend to be stable, lasting for months or even years. That doesn't mean the birds necessarily accept their position. Younger birds, in particular, try to maneuver their way up the dominance hierarchy as they grow. There's a hint of nepotism, too. In flocks of red jungle fowl (the wild ancestors of chickens), if the dominant female has a female chick, that youngster will typically join the dominance hierarchy one rung below her mother. This pushes the other birds down a level—but if they resist, they'll have the dominant mother to deal with.

This situation doesn't arise so often with domesticated chickens simply because their flocks tend to be made up of youngsters reared without their mothers. In their case, the young chickens start to jockey for social privileges at around three to six weeks of age. It's comical to see these insubstantial balls of down chasing and bumping one another and fluffing up their feathers to emphasize their size and might, but the chicks take it all very seriously. Once everyone knows his or her place and the pecking order is decided, things quiet down.

Wild red jungle fowl form flocks of up to around twenty birds, with one dominant rooster and a larger group of females and youngsters. In groups of this size, the pecking order is stable, and each fowl can recognize each of the others. In domesticated chickens, too, flocks of up to this number seem to get along pretty well, once their hierarchy is established. Often, though, commercial chickens are kept in much larger groups. If the chickens can't work out their social relationships, no pecking order is established, which can result in anarchy and ongoing aggression. Keeping chickens in large groups means fighting against the birds' natural behavior. This, in turn, forces poultry farmers to take measures to limit the aggression. One known trigger for the birds is the color red—a wound on a bird can send the rest of them into a pecking frenzy. An inventive solution is to make everything red, which can be achieved either by fitting the birds with red contact lenses (yes, really) or by keeping them under red lights. Alternatively, farmers might cut off the sharp tips of the birds' beaks.

Although the study of chickens gave us the term "pecking order," hierarchies are an important facet of life for thousands of different animals, including many other birds. The raven is a large, charismatic creature that has made a home right across the Northern Hemisphere, from Alaska to eastern Siberia. In some places they're common, but that doesn't take away from the thrill I feel when I see one of these superb animals. Recently I traveled to Thingvellir in Iceland, where the great tectonic plates on which Eurasia and

North America sit are pushed inexorably apart by immense geological forces. The landscape is primeval and craggy, yet starkly beautiful. This has been a place of execution and of an ancient parliament. If there's anywhere more raven-worthy on Earth than here, I can't imagine it. Sure enough, the birds are here in numbers. When I watched them gliding above the ominous landscape, issuing their deep, throaty *kronk-kronk* calls, they were even larger than I remembered. They have historically been thought of as animals of ill omen, the souls of the damned in bird form. As I watched, one of these glossy black birds landed on a rocky outcrop nearby and I waited in anticipation for its sonorous bass call. It undercut me by giving a high-pitched but otherwise perfect impression of a ringing mobile phone. If only I could answer and speak to it.

In fact, ravens have a broad lexicon of at least thirty different natural sounds that they use according to the circumstances. Clearly, they can augment this repertoire with some deft imitations. And far from their portentous, sinister portrayal in folklore, these are intelligent, inquisitive, and playful birds. Ravens have an intricate social structure and a hierarchy that they measure by calling to one another. Dominant ravens challenge their underlings vocally and are usually answered by a submissive call from the lower-ranking bird. When this happens, all is well and life continues as before. But if the target is feeling lucky, or cocky, it might issue a challenge, known as a *dominance reversal call*. The dominant bird can't take this lying down—the loss of face could damage its status—so the scene is set for conflict. The other ravens in the group become agitated; they know this may mean upheaval. But the extent to which they show their stress varies according to just who the protagonists are. Male ravens that hear a dominance reversal call from another male show more stress than if they hear a similar call from a female. That's because males rank higher than females in raven society, so a contest between two females doesn't disturb them so much. Females hearing a dominance reversal call originating from either sex show profound stress responses, simply because of their lower rank in the hierarchy:

it affects them more. Ravens also exhibit stress when they hear dominance reversal calls from a different group, although to a much lesser extent than when they hear these calls from their own group. This means that they build up an idea of relationships not only in their own community, but also in others, and they have a detailed understanding of their broader social environment, akin to that of the most intelligent of animals.

Ravens are members of the crow family, the corvids, which includes around 120 different species worldwide. Crows are remarkably smart and faithful. The building block of corvid social life is the breeding pair—crows, like many other birds, are monogamous, committing to their partner until death does them part. It's not just a seasonal arrangement for ravens, either. The pair remain together throughout the year, reaffirming their bond by preening one another and clasping each other's beaks in the bird equivalent of a kiss, known as *bill twining*. Living as a couple, the birds can assist one another, both in defending the territory and in dealing with squabbles among the neighbors. Raising young is much easier as a pair than on your own, as teamwork allows more food to be brought to the chicks. Moreover, if needed, one member of the pair can remain behind to protect the nest while the other goes off foraging.

For newly fledged ravens, pairing up and owning a place of their own is some way off. Having left the nest, they've made their first step to independence. But outside their parental territory, life is hard for these young birds. They offset the challenge by banding together into adolescent groups for support, in some ways rather like teenage gangs in our society. They might live in these groups for years, long enough that they develop strong and vitally important bonds with other members. Even when the inevitable happens and quarrels break out within the gang, they ultimately make up in touching displays of reconciliation, sitting alongside one another and preening each other's feathers. They come to the aid of their partners if conflict threatens, and they console them when they have been the victim of aggression. Eventually, each will leave the gang behind and begin the next phase of its life as a mature bird, defending its own territory. However, they don't forget the bonds they forged in early life, even after months or years of separation. In this, we can see the intense value of social relationships to ravens, their ability to recognize one another and to understand the emotions of others.

In the crushing cold of a Maine winter, where temperatures may drop well below freezing, ravens have to rely on their wits to find food and survive. Though the carcasses of larger animals that have succumbed to the conditions are the greatest prize, these are few and far between—ravens may have to scout for hundreds of miles to find one. Yet if they do, it represents a lifeline, a feast in a desert of snow and ice. And strangely, when a young raven finds such a treasure, it will either call loudly to recruit others, or it will make a mental note of its location and just leave, to return later with others. Within a short space of time, the carcass may play host to dozens of ravens, who strip it to the bare bones within a week. Why do they do this? After all, a single carcass could provide a raven with enough food to see it through the winter if it decided not to share. It's one thing to cooperate with close family members—evolution predicts that helping behavior should be favored among individual animals that share genes—but in such large gatherings the birds are unlikely to be related.

This was the question that drove the biologist Bernd Heinrich from the University of Vermont to spend months on end watching these majestic birds in their winter wilderness. He then described his experiences in a wonderfully evocative book, *Ravens in Winter*. Heinrich would lug huge hunks of meat, even entire carcasses, out into the ravens' realm in order to study them at close hand. Based on these experiences, he concluded that the sharing behavior of the ravens wasn't necessarily an act of selflessness. For one thing, the carcass would likely be discovered by carnivores such as coyotes, which meant that the chances of a raven being able to rely on it for food over an extended period of time were small. By sharing infor-mation, the ravens in an extended area might be able to establish a communication network that benefited all while involving little cost to the original finder, who couldn't eat it all in any case, and who couldn't in all probability keep the resource to itself over an extended period of time. In addition, a younger raven finding a carcass within the territorial boundaries of an adult pair would quickly be driven out by them. But if it recruited other juveniles, they had a good chance as a group of successfully defending the carcass from the territory holders. These juveniles had specific calls that they used to advertise their find to any other youngsters in earshot. Faced with a crowd of younger birds at a carcass, the territorial adults had little hope of forcing them all off, so had to swallow their pride and join them at the table.

PINING FOR EACH OTHER

Years ago, on a park trail near Las Vegas, I was relishing the tranquility of the moment, looking out from a high vantage point across a red stone canyon, when a flock of around a hundred bluish-gray birds came in to land near a stand of pines, shattering the serenity. They whooped and squawked with apparent delight as they rummaged along the ground for food. Noisy as they were, there was something deeply appealing about them. The birds were pinyon jays, and like ravens, albeit only a fraction of their size, they're members of the crow family. They're named for the pinyon pines, around whose annual crop of pinyons, or pine nuts, their lives revolve. Now, in autumn, they were busily stocking up, eating some but collecting far more, which they would hide away in caches to see themselves through the year. They seemed to weigh the quality of each seed like the most discerning of customers at a grocer. If it passed the test, the seed would be gulped down and added to a store in the bird's throat. As I watched, the birds crammed more and more in until their necks became distended, looking like they'd swallowed an unchewed tomato. After a few minutes of energetic but orderly collecting, they left, flying off with their bounty of pine nuts. As peace returned once more to the trail, I found myself wanting to know more about pinyon jays. Single-minded though their pursuit of pine nuts had been, there was a kind of organization to their activity, a sense of cooperation and sociability that differed from the squabbles of pigeons and many other birds as they foraged. For the next few days, I traveled in hope of seeing these engaging birds again.

In the meantime, I did some reading. I found a perfect book, *The Pinyon Jay: Behavioral Ecology of a Colonial and Cooperative Corvid*, by John Marzluff and Russell Balda. It turned out that my notion that there was something special about these birds wasn't misplaced. They're intensely social, but more than this, they live an extraordinarily coordinated life in their flock. In true crow style, a pinyon jay will form a lifelong bond with its partner. Together, the pair remain in the same flock alongside other pairs and their offspring. In fact, there can be three generations of each family unit within the flock, which as a whole could be seen as a kind of mosaic of families and clans. Although some birds, especially young females, may leave the flock to join another, to expand their dating horizons, the same birds tend to remain together throughout their lives. The intimate social relationships that build up as a result cement the flock as a fully fledged bird society.

What really struck me about pinyon jays is how the flock coordinates its activities. Every part of their lives seems to be in tune with some collective timetable, and yet no single individual apparently orchestrates this coordination. The members of the flock travel together around a large home range, searching out foraging opportunities. Once they alight on a likely spot, dominant birds usually feed first, but, having eaten, they wait politely until all have fed, so that the flock can leave together.

Feeding at ground level, they're vulnerable to predators. Thus they post sentries, often several at once, in the high branches of nearby trees. The lookouts remain still and unusually quiet for jays, taking their vigil very seriously as they watch for danger. The moment they spot something amiss, they give a warning cry to the birds below. The urgency of the danger is communicated in the pitch of the warning call—just like us, the more scared the sentry is, the higher the shriek. Forewarned, the flock takes shelter in trees where they are harder to attack. Not content with merely calling the alarm, the gallant sentry might take to the wing, following the predator and chiding it all the while, sometimes recruiting other birds to help drive off their unwelcome guest. It's such an effective defensive strategy that other bird species sometimes join the pinyon jay flock to feed under the umbrella of the sentries' protection.

While they may be aggressive when there's a need to defend the flock, pinyon jays seem fairly easygoing in their interactions with one another. After my first experience on the trail, I saw them twice more. On both occasions, I was alerted to their presence by their exuberant whooping as the flock came to collect yet more pine nuts. Arriving shortly after them, I watched as they hurried around, scooping up food. A lookout kept tabs on me, but apparently decided I wasn't a threat. Despite their constant jabbering to one another, I didn't see any aggression—the birds were too busy harvesting.

That said, this isn't a society of equals. There are social rules and hierarchies in a pinyon jay flock. The adult males are at the top of the pecking order, followed by adult females, who in turn dominate the yearlings of both sexes. Like the chickens mentioned earlier, the youngsters squabble among one another, settling their own dominance relationships within the flock. Within the group of adult males there is also competition. The high-status birds take precedence when feeding, but also don't seem to store as many seeds as their subordinates, which may mean that when it comes to the lean months, they rely on their rank to get access to the poorer birds' stores. Still, because they feed each day as a flock, rather than sneakily on their own, cheating isn't well rewarded. Regardless of status, the adults don't pick fights with the youngest birds. It's almost as though there were an unspoken rule to go easy on those in their first year of life. Despite the rivalries that occur within pinyon jay flocks, their success is based on cooperation and coordination, so it's essential in the end to stick together.

Late in the year, a change comes over the flock as new pairs form and older relationships are rekindled. As if to seal the contract, a smitten male quietly, almost shyly, passes food to his mate. Over time, this courtship feeding becomes less tender and more demanding, as females cease being coy and beg food from the males or chase them about. It takes lots of energy to make an egg, so their demands aren't unreasonable. Besides, the females want to be assured that their partners are reliable food deliverers—they'll need to be when chicks come along. A few weeks later, preparations for the breeding season go up a gear as nest building begins in earnest. While the flock still feeds together in the mornings and evenings, in the interim the pairs concentrate on family planning. Their nesting is described as colonial, but it's not the kind of colonial breeding where the adult birds crowd together to nurture the next generation. Instead, pinyon jays space themselves out, with usually only one nest per tree. What makes this a colony is how the breeding birds synchronize their behaviors. Although some will be ready to lay their eggs before others, they wait until all are ready. That they do so is another sign of how closely entrained and interdependent pinyon jay flock life is. Once the eggs are laid, each female confines herself to the nest, emerging only to stretch and attend to her toilet. While their mates are on the nest, the males shoulder the burden of collecting food, not only for themselves, but for their partner as well, and also the chicks, once they hatch.

Even though they are spread out rather than in an easily detectable cluster, the nests are vulnerable. To reduce the risk of drawing attention to the nests, the males go off to forage together, then, remarkably, coordinate their actions so that they all return with food together. The males come to their nests around once per hour, and in the best traditions of countersurveillance, rather than flying straight in, they land some way off, to check that the coast is clear—the last thing they want to do is lead a predator to their nests. If the males are satisfied, they each dart in. The male will then pass the supplies he's gathered off to his female partner with the efficiency and speed of a Formula 1 pit crew before heading out once again with the other males to collect more. As the chicks grow, the females may start to help with the feeding, but each sex has a clear role: the males feed them, while the females tend the chicks and clean the nest. Three weeks after hatching, the chicks fledge and are taken to a crèche, a kind of jay daycare. For the next month, dozens of youngsters are tended by a few adult birds while their parents go to collect food for them. It's no easy matter to locate junior amid the scrum of jostling fledglings, yet somehow they manage it. Once they've taken care of their own offspring, the parents sometimes pass food to other chicks. Though this might seem an unusual act of generosity, it also serves to keep the kids quiet. A raucous crèche full of noisy chicks all protesting their hunger is likely to draw the attentions of predators.

The amazing sociability of pinyon jays is a thread that runs through their entire lives. It begins in the nest, as siblings preen one another, and continues into the crèche, where the fledglings get a first taste of living in the wider flock. After they graduate to the main flock, the birds will continue to interact most strongly within their own cohort, with jays of their own age. Given the close relationships within the flock, it's perhaps not surprising that pinyon jays are really good at recognizing one another. Mere recognition, however, can only get you so far. In order to negotiate the complex social world that they inhabit, pinyon jays have to be able to interpret and understand not only the interactions that they're directly involved in but also those that they see occurring all around them. It's a valuable skill for animals that live in large, close-knit gatherings—each individual needs to have an idea of how it fits in. The ability of observers to assess relationships between members of their social group is a hallmark of the most sophisticated social animals. Formally, it is known as *transitive inference*. Pinyon jays, like many other highly social birds, excel at it.

But what is it exactly that makes pinyon jays so community minded? Why do they keep a lookout while the flock is feeding, or join in mobbing behavior, or synchronize their feeding and breeding, when there can be significant costs to all these activities? One possible answer is that they're helping their kin. Yet across pinyon jay flocks, the average relatedness between any two randomly picked birds is actually quite low. Another answer is that by performing these public-spirited actions, their status as a member of the flock or potential breeding partner is enhanced. There may be something in this. But the success of the flock seems to be built on a kind of tit-for-tat reciprocity—you scratch my back, I'll scratch yours.

We know that the social behavior of mammals is influenced by levels of a hormone called *oxytocin*. Humans given an oxytocin boost become more generous, while monkeys and dogs become more sociable. The bird equivalent of the hormone, *mesotocin*, seems to play a similar role in pinyon jays. While pinyon jays will share food with their neighbors on occasion, if given a boost of mesotocin they become extra-generous. A naturally occurring high level of mesotocin in the jay's system would predispose them to their flock life, to forming strong bonds with others, and could even lead them to be more cooperative.

Another thing that strikes any observer of pinyon jays is that they never seem to shut up. Their constant stream of chatter as they fed in front of me by the trail outside Las Vegas might have seemed gratuitously raucous, but communication plays an extremely important role in the group, just as it does in human groups. It enables pairs of birds to coordinate their activity as they raise their young, and in the wider flock, the vocalizations allow the jays to maintain contact with their neighbors, to identify one another, to understand each other's activities and motivations, and to warn of coming danger.

Thus far, we have made only the beginnings of understanding animal language, but we do know that many social crows have dozens of distinct calls, each of which can be given emphasis by pitch and volume. Their lexicon of calls plays a crucial role in managing their relationships and building their society. Beyond the immediate to-ing and fro-ing of vocal information, these highly intelligent birds also have excellent memories. Pinyon jays need this aptitude to recover the thousands of pine nuts that they store for food throughout the year. They also have the ability to recall past events, and to use these experiences in shaping their future behavior, including how they behave toward others. The complex relationships and societies formed by birds like these are made possible by their sophisticated behavioral repertoires, which in turn drive both the development of their supersmart bird brains and their reliance on them.

CHAPTER 5

GETTING INTO MISCHIEF

One of the least-loved mammals provides
us with a lesson in how to live.

YOU DIRTY RAT

Outside the biology department at the University of Sydney there's a patch of grass fringed with trees and furnished with a public barbecue that stands like a very Australian kind of dais. A few months ago, my lab group clustered around this splendid grill in a spirited state of mild inebriation. It was Christmas and we'd eaten, drunk, and been merry, as the season demanded. As dusk fell, squadrons of mosquitoes emerged to call time. The Sydney variety is a relentless, questing, sharp-nosed shit that can drain a human arm or leg to a flaccid, bloodless sausage skin in moments. Perhaps I exaggerate; nonetheless, they lent urgency to our ham-fisted gathering and tidying. A rustling in the newly filled bin drew our attention. A snuffling, twitching, bewhiskered nose rose into view like the periscope of a garbage submarine. Cautiously, the rest of the head emerged. A pair

149

of berry black eyes, lit by pinpricks of orange from the sodium security lamps, regarded us intently. There was a pause, almost a stand-
off, before the rat sprung from the bin with a chunk of festive sausage
in its mouth and disappeared into the shadows. All of a sudden, we
were aware of movements in the undergrowth surrounding the lawn.
It wasn't one rat, it was dozens and dozens. Moments later, the bin
was seething with them. It seemed incredible that we—a group of
biologists—hadn't noticed this encroaching rodent wave.

Maybe we shouldn't be too hard on ourselves. All over the world,
people live alongside rats, often without noticing. It was once said
that you're never more than six feet from a rat. Are we? Who knows
how many rats are out there. Living secretively, nocturnally, typically
out of sight, they're uncountable, so we have to rely on estimation.
There's no solace in the numbers for ratophobics (or, more techni-
cally, musophobics)—even conservative guesses put their population
worldwide in the billions. With the exception of Antarctica and a
smattering of small islands, they're everywhere that we are. That's no
coincidence. We create the conditions that they need to thrive, and
their adaptability has done the rest. Urban rats have evolved alongside cities since ancient times, and they grow faster and mature earlier than their country cousins. They've become savvy to the methods
we deploy to control them, and their populations have mushroomed
alongside the human ones. Much as we might resent them, their success is intertwined with our own. We made rats what they are today:
the counterpoint to our lives, the unwilling, unwanted partners of
human civilization, and we hate them for it.

When we talk about rats, we could actually be describing many
different species. The one of most interest, though, is the brown rat.
Its scientific name, *Rattus norvegicus*, is based on an age-old mis-
understanding of its origins. Though some insist on calling it the
Norway rat, it has no more connection to that country than, say, a
coconut. Rather than hailing from the fjords, the brown rat's original
home was most likely the steppes and plains of Asia, possibly in what
is now northern China. It scraped a living eating seeds and plant

matter and wasn't a conspicuous success. All of that changed when it met us. It moved in, made itself at home by our side, and took a cut of the action. Once it had formed an association with us, it spread along our trade routes and hitchhiked on our journeys to other lands. We inadvertently provided it with a ticket to ride as well as bed and board. Little wonder that it thrived, becoming not just the most successful of all the rats, but one of the dominant species on Earth.

In helpful conditions, a female rat is a pup production line. Her pregnancy lasts around three weeks, and she typically has about eight young in a litter. The pups themselves might reach maturity in as little as five or six weeks. Thus, in the space of a year, one rat can give rise to hundreds. Having unwittingly fostered and nurtured rats, our attention turns to destroying them. Yet despite our best efforts, the rats simply won't go away. Intensive rat control programs in cities using poisoned baits or gases, or even dry ice, might manage to cut their population temporarily, even by 90 percent, only for them to bounce back to their former strength within a year.

Although the appellation "rat pack" has remained in public consciousness since it was first given, as an insult, to a troop of famous entertainers in the 1960s, brown rats don't really move in packs. Instead, their furtive explorations in search of food tend to be solo. Despite this, the heart of their existence is the colony. A typical colony consists of a multitude of underground burrows and chambers, a small-scale version of the subways or sewers that the urban rats use as highways. Each burrow is shared by half a dozen or so female rats, all of whom have their own nest chamber, lined with comfortable material pilfered from their environments. There, inside their factory-like boudoirs, the females raise litter after litter of pups. The pups are born blind, hairless, and defenseless, yet less than three weeks later they're fully fledged, autonomous ratlings ready to leave the nest. Even so, the pups don't travel far to make their own home—most remain within a few yards of where they were born. So later burrows radiate out from the first, a kind of urban sprawl of rats beneath the parks and streets of our own cities.

Success for rats is more than just a numbers game. They're smart animals who can adapt to their circumstances. One enterprising rat became famous in 2015 when a YouTube video showed it carrying a slice of pizza down steps leading into the New York subway. Pizza Rat, as the rodent was imaginatively dubbed, was far from exceptional in its ability to innovate in order to grab a free lunch. Around the world, rats have adopted some amazing strategies to collect a crust. In Italy, rats have been seen diving into the Po River to collect submerged clams from the riverbed. Rats in certain parts of the United States are the bane of fish hatcheries, taking to the water to snaffle food from hungry trout, and, in some cases, even doing a spot of fishing for the trout themselves. In Germany, rats have been known to behave like miniature lions, stalking and ambushing stout little sparrows as they alight on the ground. While rats may have started as seed munchers, they are intelligent opportunists who don't pass up the chance of a broader choice of entrées.

Overlaid on their ingenuity is how they use their network of social contacts to learn from others within their society. Learning starts early for rats. Even before they're born, they pick up cues from their mother's diet, transmitted through her bloodstream. Once they're born, they show a preference for the foods that she ate, a case of mother knows best. The same principle operates when they suckle—their mother's milk is flavored by her diet and influences their food choices in later life. Rats were one of the first species to reveal this phenomenon, but they aren't the only mammals who pass on diet preferences from one generation to the next in this way. Something similar happens in us. We know that human breast milk can be tinged with a variety of flavors—vanilla, garlic, mint, carrot, and cheese all come through clearly, as do alcohol and nicotine. These are thought to create so-called flavor memories that shape our food preferences in later life.

After the pups leave the nest, they continue to learn alongside the elders of their society. By choosing to feed near adults, the young-sters share their food sources and learn as they go. Not content to be passive observers, they sometimes even snatch food from tolerant adult rats, which is one way of learning exactly what's good to eat. Although rats very often venture out alone to feed, they still manage to pass on information. Traveling between the burrow and a spot of rat fine dining, rats tend to follow walls or boundaries. This not only helps them to navigate, but also reduces the number of directions that an attack might come from. As they move along their rat run, they often keep very little distance between themselves and the wall, with the result that their fur and whiskers brush against it as they run. This marks the route with a faint odor, which can be followed and strengthened by other rats. Back at the burrow, a rat that has sampled a new food source carries on its breath the particular aroma of that food. Rats live in a rich landscape of smells and have the high-powered nasal apparatus to understand it. The stink of novel vittles on a returning forager piques their interest, even acting like an advert for that food, and influences them to try it out for themselves.

Rats are cautious animals, though—upon finding a new and exciting food source, you might reasonably expect hungry rats to pile in and gorge themselves, but they don't. Instead, they check it out carefully, eating only enough to sample it before they commit. It's this wariness that makes rats so difficult to control. If a rat eats some bait and suffers a reaction, it will studiously avoid that food thereafter. And if there's a sustained campaign of poisoning, rats become even more cautious, avoiding novel foods they come across and keeping a watchful eye on what all the other rats have been eating. They're so good at dodging the rat-catcher's bait that a thought took hold among people in the nineteenth century that, upon finding poison, rats would rush back to the nest to broadcast the news like town criers, warning their friends. Picturesque though the idea of a messenger rat is, it isn't that far from the truth, in the sense that rats collect an extraordinary amount of information from their neighbors about what's good and what's not on the local menu.

I've always been a bit of a rat fan and have kept pet rats in the past. When a cat or a dog trots out to meet visitors to a household, it will generally be greeted with varying degrees of approval. Not so when you invite them to admire your rodent. People's responses to rats tend toward revulsion. If they don't spontaneously vomit, visitors will most often edge away from the cage as though it contained a spitting cobra. Part of this is a visceral reaction to their appearance, particularly their naked, hairless tails. But as unappealing as these may be, they're a vital tool of the rat's trade. The tail stabilizes the rat as it runs and jumps and even provides a hint of grip that a fluffy tail could never equal. Perhaps a more rational dislike of these rodents is founded on the damage that wild ones cause by gnawing their way into buildings and spoiling food stores, as well as the litany of diseases they spread.

Rats have impressive teeth that grow continuously and are stupendously tough—harder than ours, harder than iron, in fact—which allows them to chew through wood and plastic. They don't need to make much of a hole to get into a building—most can squeeze through a gap of only an inch. Sometimes they don't even need to make a hole, instead finding their way into our homes by some daring, stinky navigation of the route from sewer to bathroom. The toilet thus becomes their portal into paradise. Once they're in, they bring their diseases with them—*Salmonella*, hemorrhagic fever, even the plague. Weil's disease, contracted through exposure to water contaminated by bacteria in rat urine, is one that often concerns me as I fossick around in lakes and ponds, looking for fish. My vague concerns were heightened after one of my students told me how her mother, a teacher at a Sydney school, caught it from an outdoor sink used to clean the children's paintbrushes. She went into a coma, and though she emerged from it, she had to learn to walk and talk again. Happily, she eventually made a full recovery; others have not been so lucky. As ever more people pack into cities, the contact between our two species increases, and with it, the risk of exposure to rat-borne disease.

RAT CITY

Despite sharing our cities, wild rats are notoriously difficult to study. Signs of their activities are aplenty—lozenge-shaped turds, chewed-up materials, even the slight flattening of the earth along the pathways they habitually use—but it's rarer to see them in the flesh. This is perhaps a good thing in general, but it's also problematic, because it hampers our ability to control them effectively. Though poisons might provide a temporary dip in numbers, they are no panacea. It was precisely this problem that led to one of the most famous and influential studies of animal behavior of the twentieth century.

John Calhoun spent much of his childhood in Tennessee, where his passion for studying animals first became apparent. It was a natural progression from this to university and, eventually, to a position at Johns Hopkins School of Hygiene and Public Health in Baltimore. His skills in collecting and identifying animals were put to use on the North American Census of Small Mammals, which he coordinated, as well as on a cooperative research program between the city authorities and Johns Hopkins University to understand and control rats. Even in 1946, when Calhoun began his work in Baltimore, it was already understood that you could only achieve so much with poisons, and that a broader understanding of rats and their environment was needed. With the support of the university and the freedom to lead his investigations as he saw fit, Calhoun set his imagination free. On a plot of land behind his house, in 1947 he built an enclosure that he termed Rat City. His aim was to study rat behavior at close quarters and to try to get to grips with the question of how rat populations worked. He seeded his city with five pairs of rats, providing them with everything they needed, and left them to it for a little over two years. During this time, he kept close observation from a watchtower.

Calhoun estimated that the population of Rat City could reach 5,000. He was giving the rats plenty of food, and the shelter he provided protected them from predators. And initially, they followed his predictions, thriving and rapidly populating the city. All looked well. Yet, when the numbers reached 150, strange things began to happen. The rats' behavior fundamentally changed: they went from being peaceable to hyperaggressive, and many became so traumatized that they were unable to reproduce or even to function normally. In two years the population never climbed above 200, far below the estimate. Intrigued, Calhoun replicated his experiment again and again over the next few years, with mice as well as rats. Based on the results of the first experiment, he started to frame his research in terms of population size and the effects of density on behavior. He built numerous other enclosures, each time giving the inhabitants plenty of food and structures for housing that became ever more akin to human cities, with different zones and high-rise blocks to live in. The only thing he didn't provide was limitless space. As the rodents multiplied, they started to fill the available room, and as they did so, the fabric of their societies started to disintegrate.

Time after time, what started as rat or mouse utopia became a kind of living hell. The normal patterns of interaction between neighbors broke down, to be replaced by a nightmarish chaos. Violence flourished, aggressive males attacked the vulnerable. Mothers stopped caring for their young properly, even abandoning them. Infant mortality reached a stunning 96 percent. Sex became weaponized, and mating behavior changed to the point where males would mount almost any animal they encountered. An increasing proportion of the populations—the lower-ranking, subordinate animals, in particular—showed signs of psychological trauma, clustering together in a kind of daze, alive but broken. Calhoun wrote of two kinds of death that organisms could suffer, one of the body and one of the spirit. In his overcrowded experimental enclosures, the rats and mice had experienced this second fate. Even after release from the conditions, they didn't revert back to a healthy mental state. The change was irrevocable: they couldn't function as normal mice or rats. The final stages of the experiments followed a common path—unable to operate like regular rodent societies, the populations of these rodent dystopias collapsed and died out.

Describing his experiments, Calhoun coined the term "behavioral sink" to describe the pathological effects of overcrowding on behavior. It took root in the public's imagination, especially since Calhoun seemed determined to draw parallels with a perceived decay and breakdown of human society in modern cities. Referring to his experiments in 1972, he wrote, "I shall largely speak of mice, but my thoughts are on man." The use of rats in much of his work provided a powerful metaphor, especially against the backdrop of major social changes in the United States and in Western society generally. The 1960s and 1970s saw social upheaval, protests, the war in Vietnam, all accompanied by increasing urbanization as people flooded into cities. To address the shortage of space, civic authorities crammed people into high-density, high-rise housing projects. Calhoun's work seemed to predict dire consequences. He even suggested that there was an upper limit to the number of close social interactions that an individual animal could healthily manage, and he set that number at twelve, for rats as well as for humans. Exceeding this number, he said, ran the risk of a behavioral sink, with people becoming withdrawn, even hostile.

This was fertile material for the imaginations of authors and screenwriters, who offered ominous visions of a human future, foretold by the rats, of giant urban centers with deviant, dissident, violent populations. The sense of foreboding, neatly encapsulated by Calhoun's findings, resonated in the popular press and in films such as *Soylent Green* and *Logan's Run*; in the writings of Tom Wolfe, J. G. Ballard, and Anthony Burgess; and in characters such as Judge Dredd in *2000 AD*, a British comic strip. It wasn't only in these creative fields that concerns were raised, though. Social commentators and politicians worried about the direction in which humanity was heading. Calhoun himself was more optimistic. Rather than the results of his rodent experiments predicting an inevitable and terrible future for humans, he instead suggested that they showed the need for humanity to develop creative solutions to avoid the fate suffered by the rats. His attempts at nuance were lost—after all, there's no news like bad news for capturing popular attention.

The experiments of John Calhoun represent only the tip of the iceberg in rat research. Over the past century rats have been one of the most important animals in the study of behavior. Generations of students in psychology have been reared on an academic diet of "rats and stats." These amenable rodents have provided us with a means to investigate learning, development, intelligence, play, sex, nurturing, and aggression, to name just a few. Despite our low opinion of them, we owe rats a great deal of thanks. Pioneering studies of rats have paved the way for a far greater understanding of our own behavior than we could have achieved without them.

You might well ask why we study rats to understand ourselves, and it would be a fair question. Rats are among a select group of species that we refer to as *model organisms*, animals (or plants) that we study as a means to answer questions and unpick conundrums in biology. Rats are, like us, mammals, meaning that we have much in common with them. Our bodies are built and run in similar ways; our motivations and our responses to stimuli also have many similarities. Of course, as John Calhoun learned, you can't always extrapolate directly from the behavior of rats to that of people—there are some rather important differences between us. Nevertheless, you can build a framework of understanding that can give you a starting point for detailed investigations of human subjects. An example can be seen in the transmission of food preferences from mother to offspring, but there are many more.

Calhoun's experiments mark out rats as feisty, aggressive animals, yet that possibly says more about the breakdown in their society under the influence of crowding than about rats under normal circumstances. Rats, like us, and like so many other social animals, spend much of their lives among a coterie of friends, relations, and neighbors. Interacting with the same individuals over a period of time puts a premium on good behavior, or at least on a kind of "you scratch my back, I'll scratch yours" trading of favors. It shouldn't surprise us, then, that rats can be highly cooperative, that they don't behave like, well, rats. Even so, the sophistication they show in deciding whom to help, and when, is remarkable. Sometimes, their helpfulness is born out of a sort of feel-good factor, a contagious kindness that comes from having themselves been helped out in the recent past. In our own society, we often see this same cooperative domino effect—for instance, during rush hour driving, when someone lets us out of a busy junction in traffic, we're more likely to be considerate to the next person we see waiting to pull out. It doesn't just help them, it makes us feel good, too. That said, neither we nor rats are mugs; we each often tailor our efforts to be nice to someone according to our past experience of the individual in question. Rats provide more help to those who've recently done them a good turn than to those who've recently screwed them over.

Reciprocal altruism, as this behavior is formally known, promotes cooperation between unrelated individuals in animal societies, particularly those where a group of individuals live side by side. Each animal might aid a neighbor on the basis of having been helped in the past, or in the expectation of a future benefit. At the same time, selfish behavior gets filtered out: within a stable social scene, being a jerk is a bad strategy in the long run. But despite the apparent all-around wonderfulness of the concept, reciprocal altruism isn't all that common in animals. What can look on the surface like cooperation can actually be a powerful, high-ranking individual exploiting a weaker one's eagerness to please; a kindly act might disguise an ulterior motive. On top of this, the payoff for taking a free ride seems to be irresistible to some, especially those who are just passing through, because they won't have to reap the costs of their antisocial actions. In human society, social norms can serve to reinforce a moral code of behavior—disapproval can be a powerful means for keeping us on the straight and narrow. How many of us can recall when as children we withered inwardly when a parent deployed the nuclear option phrase: "I'm not angry with you, I'm just disappointed . . ."?

Do animals understand or shape their behavior according to some set of social rules that defines what's expected of them? Maybe some do. Honestly, we don't know. Nonetheless, rats—and especially female rats, who are more sociable than their male counterparts—do seem to be highly cooperative. What's interesting is how rats tailor their cooperation to the circumstances. For instance, they're more likely to donate food to a hungry neighbor than to a less hungry one—which seems to indicate that they provide help according to their partner's needs. That said, hungry rats aren't shy about asking for food—they beseech potential donors to provide it for them, reaching out and calling to them. We don't necessarily hear the pleading of hungry rats because much of their vocal communication is in a pitch too high for us to pick up. In some circumstances, hungry rats are also more cooperative themselves, which suggests an awareness of not only their own plight, but the economics of the rat trading system.

Though food is a vital currency in these exchanges, it isn't the only commodity that rats deal in—being on the receiving end of a nice bit of grooming encourages rats to give food to their rat masseuse next time the need arises, or to reciprocate with some grooming efforts of their own. Within this system, good character goes deeper than simply being willing to offer food. In deciding whether to pass morsels to a recipient, rats take into account that recipient's recent behavior toward them. An aggressive rat may face being cut out of the cooperative network, so it pays to be a good citizen. Despite rats being a byword for devious immorality, the truth is that they show a sophisticated level of cooperation geared toward rewarding helpful behavior.

Although negotiating the tit-for-tat networks is a vital part of rat society, the closest relationship is that between a mother and her pups. Indeed, a strong bond between mother and infant is almost universal in mammals, and of course that includes us. Providing mewling newborns with milk isn't enough on its own to set them up for healthy development; physical and emotional maternal closeness during early life is a crucial factor. We've known for a long time that a tough childhood can mark a person for life. Sadly, our societies are rife with those whose adult anxiety can be traced to a traumatic upbringing, and our prison populations are disproportionately made up of people who suffered in their early years.

Strange as it might seem, it was studies on rats that first shed light on just why this is. Among rats, there are some who make excellent mothers, diligently nurturing, grooming, and nursing their pups. And then there are those who apparently believe in the idea of tough love, and are neglectful of their offspring. These differing experiences have profound effects on the pups. Those who benefited from the close attentions of a caring mother grow up to be calm, well-adjusted rats, whereas the pups of lax mothers develop into anxious adults. If that weren't enough, those anxious adults occupy places low down in the rat hierarchy and are more likely than their emotionally healthy peers to suffer serious diseases. It's not all bad news for the anxious rats—in a world full of danger, it can be useful to be watchful and jumpy—but it seems scant consolation.

We're used to thinking of genetics as the conductor of our developmental orchestra. Genes are made up of the DNA that is sometimes referred to as the instruction manual on how to build a body. Implied within this is the idea of genetics dictating a rigid, unbending destiny for a growing organism. But although genes are important, they are not all-powerful. The immediate environment of a developing animal plays a massive role in determining if and when genes get switched on; in other words, it allows the flexibility to shape a growing individual to fit the world in which it will live. This is where the mother rat's behavior plays such an important role. The attentions of a nurturing mother increase the expression of a gene that ultimately helps the pup deal with stress.

That's rats; what about us? There's tentative evidence to suggest something very similar goes on in human babies—those who are breastfed and cuddled more show patterns of gene expression similar to those of nurtured rat pups. Fortunately for those who might now be panicking about whether their children, or indeed their pet rat, got enough love in early life, the good news is that the effects of any early shortcomings are potentially reversible. Rodents that suffer the inattention of wayward mothers gradually develop an improved response to stress if they experience pleasant environmental conditions as they grow. Obviously, rat pups aren't the same as children, but it doesn't seem too controversial to suggest that a child who is nurtured and stimulated has a better chance of becoming a happy adult than a child who is neglected or abused. Thanks to pioneering research on rats, we now understand just why this might be so.

As rat pups emerge from their mother's nest to enter into the wider world of their community, they start to mix with their peers, at which point they are exposed to a gamut of scary new experiences. The presence of other rats makes a fundamental and largely beneficial difference to how they deal with these. On the one hand, being around other rats provides the pup with the chance to learn about life's dangers secondhand, without having to take them on directly, and possibly fatally. On the other, the company of other rats is in itself a balm for a stressed rat's nerves. These communal benefits aren't exclusive to rats—we see them in all manner of social animals, from fish to people—and they play a crucial role in the success of such animals. Nevertheless, it is rats that have provided us with our detailed insight into how communal benefits work.

Every day is a school day for a young rat. Watching the other members of its society going about their business, it has an excellent opportunity to glean info on the good things and the bad things in life. It learns, as we've already seen, about tasty and safe food. It also absorbs evidence about local perils and pitfalls. When rats see one of their number react with fright to something or make a terrible mistake, they internalize this and adapt their behavior accordingly. They don't have to see things for themselves to go on red alert—a frightened rat transmits its fear to others. This isn't necessarily communicated as directly as "Guys, you know the cat that got Reggie? It's outside and it's got that look about it," but nonetheless it's clear to other rats when one of their number is stressed. An anxious rat manifests changes in its behavior and emits a particular smell, instinctively recognized by its neighbors, which tells of its fear.

This sense of dread passes like a contagion among the other rats, whose heart rates speed up. They become more wary lest they, too, run into trouble. The response is strongest when they're able to link the other rat's fearfulness with the thing that provoked its reaction—perhaps some strange, predatory animal, or a bare wire that gave the rat a shock as it nibbled at it. The others gain a specific understanding of the cause and effect, and they learn to avoid the source of their rat colleague's distress. Incidentally, people also give off a particular odor when they're scared. The reason we aren't more aware of it is that our sense of smell is exceptionally poor compared to lots of other animals. Nonetheless, the chemicals we produce when we're frightened are expressed in our sweat. Dogs—and lots of other creatures—really can smell our fear. Some people can distinguish between the odors of people who've watched a horror film and those who haven't, or between the sweat of novice skydivers and that of gym-goers. Women are particularly good at this, which reflects the fact that, on average, they have a better sense of smell than men.

Valuable though emotional contagion can be to social animals, if unchecked it can cause havoc and give rise to mass hysteria. In the Middle Ages, dancing manias spread throughout Europe, affecting swaths of people, who would dance madly for hours, sometimes in groups of hundreds, until they were overtaken by exhaustion or even died. Nunneries were the unusual location of many outbreaks. In some cases, groups of nuns became very un-nunlike, cursing and acting in a sexually provocative way; in one instance, the nuns all started to meow like cats. It being the Middle Ages, the finger of blame was pointed firmly at the devil. Certainly, it represented good business for exorcists and holy men, even though they proved to be a bit hopeless in quelling the hysteria. In recent times, there have been multiple recorded outbreaks of mass fainting, hysterical screaming, or uncontrollable epidemics of laughter in schools and factories around the world. Often these can be attributed to gatherings of closely connected people being in a heightened emotional state— what starts in one person can spread rapidly, tipping large groups into hysteria. There are animal equivalents of these outbreaks, but they're surprisingly rare—there are a few instances of flocks of broiler chickens in large growing sheds descending into a collective panic, for example, while a large group of baboons at a zoo in the Netherlands abandoned all their usual monkey business to sit unresponsively in a group huddle in 2013.

The idea of runaway emotional contagion and mass hysteria is something we're all aware of, yet at the same time our streets aren't often full of manic dancers; nor is the natural world beset with the problems of mass fainting in herds of wildebeests, or anything like it. Aside from the rather sparse collection of anecdotes that I've given, examples are few and far between. Even though lots of social animals pick up on the distress or fear of other group members, it doesn't escalate out of control.

This is interesting in itself—it suggests that something acts to keep emotional contagion of fear in check. And indeed, there is something at work: it's known as social buffering, and it's hugely important to group-living animals. Simply put, social animals benefit from the calming influence of others of their kind. Social buffering reduces anxiety and stress, sometimes dramatically, and accelerates the recovery of the original nervous animal. In rats, close contact with the mother is the norm from an early age. The pups, born helpless and hairless, huddle together alongside her for warmth. Even after they leave the nest, the physical presence of other rats in their community buffers them against the cares and concerns of the world. Although a lonesome rat will gain some benefit from being with a stranger, the effect is strongest when the rats share an affiliation. Although we might say rats seek out their closest companions to calm themselves, it might actually be better to say that rats seek out those who most calm them and choose these as close companions. It's a subtle difference, but potentially an important one, one that very probably shapes networks of relationships in all social animals, including us. Aside from the sights, sounds, and touch of nearby animals within the hubbub of the rat colony, even the smell of a happy-go-lucky chum can calm an edgy rodent. We know that in rats these social stimuli suppress activity in a part of the brain that fires up the fear response, so when we talk about social buffering, we're not describing a vague sense of calm and well-being, but a direct influence on the animal's physiology and the way it thinks.

Social buffering plays a major role in our human societies. For babies and children, it's the parents, and especially the mother, whose presence gives them this stress release in the brain. Interestingly, although a pacifier hushes a baby, it doesn't quiet the anxiousness going on inside the baby's brain to the extent that mum's company does. As we get older, our social peers take on more and more of this role. That said, it's not as simple as saying that friends take over the social buffering role from parents. Instead, the attachments formed by a young child to its parents in early life provide the foundation for that child's ability to form close friendships and so to benefit from the stress release of social buffering as they mature into adults.

One of the psychologists' favorite tests of anxiety in people is known as the Trier Social Stress Test. In this test, subjects are asked to prepare and give a short speech to an audience and then do some mental math. These tasks very reliably and consistently cause stress—you're not on your own with this—and that's of course what makes it a good test. To understand just how the Trier Test affects people, we can then measure levels of a hormone called *cortisol* in the saliva, as well as heart rate, both of which indicate levels of anxiety. After that, we can see how the presence of a parent, a friend, or a partner buffers against the stress. Parents diffuse the tension for children in these situations right up to, and sometimes past, adolescence. With friends, the picture is more complicated—it can work both ways, especially for adolescents. While close friends might reduce stress for young adults in these tests, for some of the test subjects having friends or peers nearby can actually increase the level of stress while they do the tests. It really isn't easy being a teenager. For adults, things stabilize a little, although there are still some strange—and as yet not fully explained—differences between the sexes. Heterosexual men seem to get a bigger stress-relieving boost from their female partners when preparing for the Trier Test than the other way around. It might be that while women generally support their male partners in this situation, some men manage (even if unintentionally) to undermine their female partners. This may have to do with how the sexes interact while preparing for the test, and particularly with how they discuss the coming task. In different tests that are run on women, while they are holding hands with their male significant other in silence, or even being massaged by them, women's anxiety does show a reduction. Perhaps the message to men is to shut up and simply "be there." There's a long way to go with this kind of research before we can join up the dots and really understand what makes us work, but again, studies of rats provided the groundwork for working out how social animals tick.

If animals can transmit both anxiety and calmness to one another, we might envisage a battle for supremacy between these two contrasting forces. In the red corner, the frightened rat, giving off all kinds of fear-promoting signals and smells. In the blue corner, the calm rat, with its aura of reassurance. Who wins? Does the calm rat become frightened, or does the frightened rat become calm? It's a fine balance. Rats are complicated creatures with a great deal of variation between them, not little computer programs that give a consistent answer each time. Certainly a stressed rat will become calmer in company than it would be on its own. Going in the other direction, although the stressed rat might be sending out signals of its anxiety, its mere presence will act as a social buffer for the calm rat. Even so, the calm rat will likely become more anxious than it was; the nervousness of the other rat tells the calm rat that it needs to be watchful. If the calm rat remains blasé, indifferent to the possibility of danger signaled by its colleague, it will probably end up sooner or later as a dead calm rat. Ultimately, the spread of fear from one rat to another is dampened and buffered by the presence of other rats, but not wiped out entirely. That's why wariness increases, while not giving way to a mass panic within the rat colony.

While rats benefit from the social support provided by the presence of other rats, the same couldn't be said of their interactions with us. I remember a rodent outbreak in our outdoor shed when I was a teenager. One particular rat sat calmly grooming itself on the roof in broad daylight. My dad grabbed a sledgehammer and tried to sneak up on it. The only real danger to the rat was if it laughed so much it choked. As the great hunter approached, the would-be victim nonchalantly finished its hygiene routine before sauntering off to warn the others that it'd seen an idiot. My dad isn't an idiot,

I'm just putting words in the rat's mouth, but his response to seeing the rat is a common one. People react strongly to rats, and it's understandable in the light of the damage they cause and the disease they spread. Yet for all that, I think we have to admire them, even if grudgingly. They're smart and innovative, and they cooperate with one another. This much I already knew; yet I have to admit that even I was surprised by some recent studies that show rats in a completely new light.

Imagine for a moment that you're at home looking out the window on a miserable rainy day. You see a person outside, soaking wet through. What do you do? Invite them in to dry off? Some would, others might just draw the curtains and go back to their book or PlayStation or whatever passes for wet-weather entertainment in their house.

Rats, it turns out, are door-openers. In a recent experiment, pairs of rats were housed in adjacent quarters. One living space was dry and pleasant; the other was wet and less pleasant. Though rats can swim, by and large they'd rather not. Connecting the two areas was a door that could only be opened by the rat in the dry housing. The result is two rats, one bedraggled wet rat and a dry rat with a decision to make: Should it open the door to let the other in? The dry rats in the experiment did, and what's more, they were quicker to open the door if they'd suffered in the same situation themselves. This result seems to suggest that they can identify with the discomfort of another rat and that they're quick to lend a paw to help it out. In a similar situation, minus the water, the rats didn't open the door. In other words, their motivation to let the neighbor in was based on the other's needs, rather than the dry rat's own desire for company.

In another test, rats were confronted by the presence of a fellow rat that was trapped. They had the option to free it, or not to. Overwhelmingly, they chose to free it. Why? They didn't release the trap if there wasn't a rat in it, so it wasn't simply a matter of abstractedly playing with the equipment. Instead, they seemed to recognize the plight of the incarcerated rat and to decide to help it out. Empathizing—the ability to identify with another's emotional state—is something we take for granted in ourselves, but rats? Dirty, thieving, stupid rats? Well, yes, they seem to feel it too. So determined were they to help out, in fact, that they passed up the chance to chow down on some chocolate, a rat favorite, in order to assist a rodent in need.

There's been a good deal of debate and criticism of these studies, in particular about the conclusion that rats open the door, or free the trapped rat, because they're motivated to help when they see a fellow rat in distress. A simpler explanation is that the rats who helped did so because they wanted another rat to hang out with, which means their motivations were entirely selfish. Getting inside an animal's mind to untangle the mystery of exactly why they do something is a massive challenge. The question of empathy is important, though, because it sheds light on our evolutionary origins; it can tell us about the roots of our humanity, our society. That's the big picture, and we can learn something else from this, too, if we look at how stress affects this display of altruism.

I've already described how social animals buffer one another against day-to-day stress. Though we're used to thinking of stress in exclusively negative terms, actually a reasonable amount of it can be a good thing. When I'm heading into a lecture hall to, I hope, engage students with a dazzling constellation of biological wonders, a little adrenaline makes me do a better job. Too much, as was the case when I gave my first lectures years ago, and I stumble over my words and get lost in mental cul-de-sacs. Too little, and I become vapid and bland and don't hold my audience. There's a sweet spot where I'm just anxious enough to perform at my best. Countless people in the public realm, from politicians to sportspeople, have said something similar.

What has that got to do with the workings of animal societies? It turns out that a moderate amount of stress promotes good behavior in rats. They're more likely to seek out others and build stronger relationships with them if they are just a little stressed, with the result that they act less aggressively and share their resources. They become more helpful to one another. Totally calm, stress-free rats are less motivated to involve themselves with the nitty-gritty of communal living—they don't need the social support to buffer their anxieties. Highly stressed rats are also less engaged than the moderately stressed ones, though for different reasons. Severe upheavals or threats can cause them to withdraw into their own psychological world. They struggle to make and to maintain social ties in their rat networks, and so they become isolated, showing the rat equivalent of human depression, or even post-traumatic stress disorder.

Although we obviously have to be cautious about extrapolating from rats to humans, there's a lot of overlap between the basic physiology of a stressed rat and that of a stressed human. It's not breaking news that large amounts of stress can be damaging to people, although it might be surprising to learn that rats are susceptible to it, too. Yet seeking to medicate it out of our lives altogether could have detrimental effects for us as social animals. In this area, as with so many others, rats have given us great insights. Maybe, just maybe, when we see one of these complex yet detested creatures, we shouldn't automatically reach for a sledgehammer.

BARE-FACED RATS

The inspiration that many people feel when amid the glory of nature was captured perfectly by Charles Darwin in the immortal phrase "endless forms most beautiful and most wonderful" (it's how he ended his book *On the Origin of Species*). But while there are countless creatures capable of elevating the human mind to poetry, there are a fair number that fail to set our hearts racing in quite the same way, among them the rat. Even so, one type of rat in particular stands out for its ability to make us gasp at its sheer hideousness. This is the naked mole-rat, a distant cousin of the brown rat and an animal that was described to me by one of the researchers who studies them as a penis with teeth. When I first came face to face with them, at a university in western Germany, this wasn't the first thing that came to mind. Once an idea is planted, though . . . And so I found myself watching these gnashing little phalluses scurrying around their habitat, simultaneously appalled and fascinated by their singular appearance. Not that you would call them ugly to their face—I was told that the largest ones could do some serious damage, though, in the lab, usually only with things like carrots, which they can bite in half with a single snap of their buck teeth. So it was easy to decline the offer to hold one. This isn't an animal you'd want to dandle on your knee.

But though they may be the most left-swipe-worthy animal in the biological catalog, beauty is in the eye of the beholder: for scientists studying naked mole-rats, there's much to admire. Take, for instance, the fact that they can live into their thirties—the brown rat, by comparison, is larger, yet typically lives for only about one year. Naked mole-rats can survive at low oxygen levels that would kill us, they're almost completely immune to cancer, and they have no pain sensation in their skin. They are extraordinary in so many ways that once we look past their desperately unfortunate appearance, there are myriad things that we can learn from them. Not least of which is the way that their bodies clamp down on cancer cells, using a chemical known as *hyaluronan* that shows potential for use in human medicine. That said, what makes naked mole-rats most interesting to behaviorists is their lifestyle.

In extensive and expertly constructed burrows under the dry soils of East Africa, societies of naked mole-rats live in a manner unique to mammals. They've taken a leaf from the social insects' book and have adopted their winning formula. Like ants, bees, and termites, mole-rats are *eusocial*—their groups represent the pinnacle of social organization. Within their colony, there's one mother, the queen, a single female with the right to breed. She has a small and select number of male consorts, but beyond this little group of idle lotus-eaters, the rest of the colony's members—numbering into the hundreds— are sterile, indentured workers. Their sterility is a means of control, enforced on the workers by the queen, who is a dominant figure in the colony. Take the workers out of her oppressive orbit, and they become fertile, though few ever escape her feudal regime. The smaller mole-rat workers have the task of roaming the maze of tunnels in search of roots and tubers to feed the whole group, while the larger ones are given the job of colony defense. At the heart of the colony is a single sleeping chamber, where all the mole-rats gather after a hard day or night (it scarcely matters underground) of fetching and carrying. In this communal bedroom, the whole mass of rodents huddles together—royalty, workers, and pups alike. They also have a larder chamber and a toilet. There's quite a bit of overlap between the larder and the toilet, in that mole-rats are avid consumers of turds. The young get fed on them once they're weaned off the queen's milk, and the workers, too, chow down on them. Eating and then re-eating is a very efficient, if not particularly appealing, way of making sure you get the most out of any food. Not only that, but eating a queenly crap, marinated in her hormones as it passed through her regal body, primes the workers with estrogen and induces them to be better surrogate parents to the queen's pups.

Reassuring though it may be to spend decades in the same burrow, with largely the same group of animals, there's a small number of naked mole-rats who dream of bigger things, and of the world beyond. It's just as well that some do, because the genetic penalties of endless inbreeding can be extreme. Despite their incipient wanderlust, the dispersers are often spectacularly lazy while still inhabitants of their birth colony. They don't engage with the busy rota of mole-rat business, instead hanging back in studied indolence, packing on fat in readiness for their great escape.

Above ground, however, the world is full of peril—there are plenty of predators out there who are hungry enough even to eat a mole-rat. Their chances of reaching another colony and becoming a breeder there, or of founding their own colony, are small. Nonetheless, some do make it, and once they leave the controlling influence of their queen, they are able to shrug off the enforced sterility of the worker and finally reach maturity. Sometimes a small posse will leave together; at other times, a loner will seek to forge its own way in the world. When this happens, it will dig a small burrow and festoon the soil around the opening with poo cues that serve as a "come and get me" message to the next adventurer that happens along. By such unorthodox means, naked mole-rats spread to colonize pastures new.

There are lots of potential answers to the challenges that life poses. Brown rats hoisted their banner next to ours, while retaining the close company of other rats as a buffer against the everyday struggle that comes with being a small, vulnerable animal. Mole-rats have taken sociality a step further, sacrificing some of their individuality to intertwine their existence with those in their colony. Although neither is ever likely to win the hearts and minds of humanity, both are incredible, successful social animals nonetheless.

CHAPTER 6

FOLLOWING THE HERD

Close ties among herding animals provide the foundations of empathy.

DOWN ON THE FARM

When I was in my early teens, my family moved out of town and into the countryside. For me, this meant more creatures to look at, which was good news. And since I was now surrounded by farmland, it also meant exposure to an entirely new palette of smells—fifty shades of bouquet, if you like. Most of these were shitty, in one way or another, but that's what goes with living near a farm. I didn't give much thought to the farm animals themselves, beyond a kind of frustration while playing soccer with friends in the fields when, setting myself up for a perfect half volley, the ball either stopped dead in the glutinous embrace of a cow pat, or ricocheted giddily off a sheep turd. The animals looked on with total indifference to the ruination of my sporting prospects. In fact, they seemed to look at absolutely everything with indifference. Even for a zoological enthusiast like me, they just blended into the background. As time went by, however, I started to see the subtleties in their behavior that my casual, initial appraisal had missed. Rather than being insensate lumps of mobile meat, they were individuals with quirks and personalities. As with almost everything, the more you learn, the more you develop an appreciation.

Let's start at the beginning. The beginning, in this case, being the transition of humans from hunter-gatherers to pastoralists and farmers after the most recent ice age. This agricultural, or Neolithic, revolution was a big deal. There's considerable dispute about when our own species first emerged, but a reasonable estimate is three hundred thousand years ago. Yet it was only around twelve thousand years ago that we began to harness nature in order to grow crops and to rear our own animals. With this change came another, this time in the way that we lived. We gradually gave up the itinerant lifestyle and became sedentary, building settlements in which to live. We became cultivators, harvesters, and herders. Our society changed as we began living in larger, more concentrated groups.

Across the world, there are something like thirty major species of domesticated mammalian livestock. All of these are herding animals—they all show social behavior. This is no accident; it's pretty much a prerequisite. Darwin recognized this, writing on the topic a century and a half ago: "Complete subjugation generally depends on an animal being social in its habits, and on receiving man as the chief of the herd or family." Animals that live in large groups lend themselves to domestication, both because they're clustered in the landscape and because they move—and so can be herded—together. The epicenter for the Neolithic Revolution was the so-called Fertile Crescent, an arc of land rising from the banks of the Nile and running up the Mediterranean coast to modern-day Turkey, and then curving back south to follow the courses of the Tigris and Euphrates Rivers to the Persian Gulf. It was in this Fertile Crescent, the cradle of civilization, that we humans made one of the greatest transitions in our history. A combination of rich soils and a helpful climate supported the cultivation of crops and the growth of lush grasses to support animals.

All of what you might call the "big four" livestock mammals were first domesticated in the Fertile Crescent between ten thousand and twelve thousand years ago. Modern sheep come from Asiatic mouflon in what is now Iraq. Goats, from ibex, and pigs, from wild boar, also began their journey in the same region. Cattle, too. Astoundingly, evidence from DNA points to domesticated cattle emerging from a single herd of just eighty wild ox, or aurochs. Other forms of cattle, such as the zebu, with its characteristic humped back and floppy ears, are thought to have been domesticated from local aurochs populations in different parts of the world. Even so, as cattle have spread and interbred, the vast majority of the world's billion-plus cattle can trace their origin to the original eighty wild founders in the Fertile Crescent. The early farmers could only work with the animals that lived around them. Had the local fauna been different, or had the Neolithic Revolution happened in a different part of the globe, who knows what farm animals we might have now?

As early farmers spread from this agricultural hotspot, they took their animals with them. Our idea of livestock farming usually incorporates fences or walls to keep the animals in, but that's a relatively modern development. For most of the history of farming, keeping the animals together has been a challenge, and shepherds and cowherds really earned their corn. It also meant that as domestic animals crossed paths with their wild counterparts, the wild animals might be incorporated into the herd and the animals might interbreed. The wild aurochs, mother of all cows, became extinct in the early seventeenth century (the very last bull aurochs had its horns made into drinking vessels for King Sigismund III of Poland; thus noble beast became novelty tableware). Yet the aurochs lives on genetically in modern cattle. This is especially true in the British Isles, where DNA analysis shows that local aurochs were on particularly friendly terms with their domesticated counterparts until medieval times. The finding has encouraged some to dream of reclaiming Europe's lost

biological heritage, back-breeding domestic cattle to restore the aurochs to life. Or, rather, something like the aurochs—it's not possible
to piece together fragments of surviving DNA to create an entirely
faithful simulacrum of an extinct animal. The motivation for this
endeavor is to rewild parts of Europe and rebuild a historical ecosystem. The aurochs was once a major part of the region, shaping the
landscape through its activities. One of the roles of large grazers is
to control the growth of trees in unmanaged habitat—without grazers, you end up with woodland alone. By including "Aurochs 2.0,"
you might foster the development of an authentic, natural mosaic of
meadows and woods. That's the idea, anyway. Competing projects
in the Netherlands, Germany, and Hungary are already underway,
and the quest to restore areas of European wilderness on abandoned
farmlands is gathering pace.

Before cattle and the others were brought to live in close association with people, they were hunted as wild game, so likely had a
healthy aversion to us. Though we can only speculate on the process
of domestication, a reasonable scenario is that the switch from hunting to farming began with the hunters applying strategies to maintain
the supply of animals, including making sure that competing carnivores didn't snaffle what they would have considered their own food.
Over time, it gradually involved exerting some control over where the
animals went, until people ended up herding them. It sounds simple,
but wild animals, then as now, are unlikely to greet the approach of
a hunter with unbridled delight. The transition from skittish, jumpy
animals that would flee at the first sign of a spear being sharpened to
the pliant denizens of modern agriculture took time. Moreover, an
animal like an aurochs, even a wild boar, is a substantial and dangerous proposition. The aurochs could be huge, able to look a tall man
in the eye and possessed of wicked, 3-foot-long horns. In his account
of the Gallic Wars, Julius Caesar described aurochs with admiration:
"Their strength and speed are extraordinary; they spare neither man
nor wild beast which they have espied." Still, individual animals
vary in their docility and their propensity to be tamed, and it was

probably the most easygoing that were most amenable to a shift in their living arrangements. Even today, the most difficult and truculent animals are routinely chosen to be slaughtered first. With a modern understanding of genetics, it's easy to see how this practice selects for particular behaviors. In those ancient days, people had little or no concept of how to manipulate traits; nonetheless, the unplanned actions of early farmers in favoring or finishing off particular animals had the net effect of causing a gradual shift toward breeding more compliant, manageable livestock.

Although deliberate linebreeding of animals to develop particular strains is something that has a fairly short history, thousands of years of domestication have produced substantial changes in farm animal behavior. Compared to their wild forebears, today's farm animals are tamer, less active, and less aggressive. They also have smaller brains—substantially so, in some cases—with up to a third less gray matter between the ears than their ancestors. Cattle have only around half the amount of brains you'd expect in a mammal of their size, while a pig has less even than that. Does that mean we've bred them to be stupid? Not so fast. What we *have* done is to breed them to be disproportionately heavy and meaty. Goats and sheep, which haven't undergone the same intense selection for meatiness—they're less fleshed out—have brains roughly the right ratio to their body size. Another aspect of livestock neuroscience is that nowadays we often rear animals in dismally understimulating industrial conditions. The brains of animals raised in this way show deficits in size and complexity compared to those that have had the benefit of a rewarding and complex environment. Nevertheless, farm animals are not stupid, however much we might like to pretend they are to comfort ourselves when we eat them.

A tweet by the physicist Neil deGrasse Tyson a couple of years ago framed what many people think about cattle: "A cow is a biological machine invented by humans to turn grass into steak." It was a rather bare statement, and one that, predictably, caused an awful lot of pushback, as well as, it must be said, a good deal of agreement.

The question remains, is he right? In some ways, he is. The problem isn't so much with the logic of it, but the reductionism. His tweet redefines a living, sentient animal as a mere object. It's perfectly reasonable, if a little cold, to say that you don't care about cows. It's unreasonable to say that they aren't intelligent animals with the capacity to suffer.

Take, for instance, their ability to identify one another. Being part of a social group means developing and maintaining relationships, which depends on recognition. It's perhaps no surprise, then, that cattle can tell one another apart. What is surprising is that they can do this from portraits of other cows' faces—from a simple two-dimensional image. Facial recognition is, of course, something that comes naturally to us, yet it's not a simple matter. The way we perceive a face changes both according to the conditions—the angle we see it from, the lighting, and whether the person is moving—as well as the emotion and expression of the face. Our brains don't store a fixed image of a person's face; that's why, if you try to picture someone's face in your mind's eye, most people come up with something fairly fuzzy. Despite this, when you see the face of a close friend in a crowd, you can pick them out instantly. How? Your brain has specific and highly specialized areas devoted to facial recognition. Within these areas, distinct brain cells each focus on one particular attribute of a face. For instance, some cells concentrate on the size and relative position of the nose, others on the shape of the lips, while still more attend to the distance between the eyes, and so on. Essentially, you don't memorize a whole face in a single image; rather, you remember *parts* of faces. For each person you see, different parts of your brain collaborate to create and recognize an identikit mosaic of facial characteristics. Incredibly, by studying the detailed activation patterns of cells in the brain of a primate looking at a face, it's now possible to build an image of the face they're seeing.

All of which is to say, facial recognition is complex, yet cows can identify each other from pictures. They can also distinguish between people, even if those people are dressed identically. They recognize and remember people who've treated them badly and, conversely, those who come bearing gifts in the shape of tasty morsels. Sheep, too, can do this, perhaps even better than cows. They can identify at least fifty different members of their own flock by their faces and can even reliably recall those faces after two years of separation. Moreover, seeing a familiar face when they're uneasy, such as when they're isolated, calms them down: their heart rate slows and the levels of stress hormones in their blood decrease. Like cows, sheep can tell us apart based on our mugshots. Funnily enough, they can also pick out particular people, those they're most familiar with, from lineups. Both sheep and cows are better at recognizing members of their own species than they are at telling people apart—it's what their brains have evolved to do—yet the fact that they can perform these tasks so well is a clear indication that they're far from the dumb animals that many people take them to be.

That they are able to recognize one another from pictures gives us an insight into what you might call their *social cognition*, which is all about navigating relationships within the herd or the flock. But another aspect to intelligence is how individual animals relate to their environment and find their way around. In the wild, they need to know where food or water can be found, and where to find shelter if danger threatens. You might reasonably argue that a cow in a field doesn't need to know where the grass is (it's standing on it) or where shelter is (because the farmer will take them there). Fair enough. But millions of years of evolution are not extinguished entirely by a few millennia of domestication, especially when you consider that the fences-and-fields lifestyle of modern cattle has existed for only the past couple of centuries.

One way of testing spatial learning is to use a maze. There are standardized approaches to this that let us scale the task to the size of the animal and so compare between species. The animals learn that they have to solve the maze to get a reward, such as a bucket of delectable cow chow. The tricky part is that there are often a dozen different maze configurations to learn. When the test comes, they're presented with one of these twelve at random, and they have to think back and remember how to crack that particular labyrinth in order to get their reward. It turns out that cows are pretty good at this particular form of mental dexterity—in one such test, they even did better than rats and domestic cats. Again, not bad for a mere biological machine.

Maybe Tyson wasn't really thinking of the intelligence of cows when he tweeted his thoughts about them. Maybe instead he was thinking about their lack of expression, their vacuous appearance. It's true that, at first glance, cattle don't show much emotion, and that this makes it harder for us to identify with them. But it's a feature that's common among lots of animals that have evolved under the constant threat of danger. It's sometimes said in our society that convicts entering prison for the first time need to avoid showing weakness or else they become targets. After this initial experience, inmates often develop what's known as a prison mask as a means of concealing their innermost feelings. There's some similarity here with prey animals—they both have to present a front. For prey animals, expressions of pain or vulnerability are a signal to predators on the lookout for an easy victim. Wild cattle, and the ancestors of our other farm familiars, shared their habitat with a host of carnivores. Any individual making a fuss would invite attention. Hence, their descendants come preprogrammed to mask their weaknesses. Cattle are very British about showing emotions; they're stoic in the face of suffering. Nonetheless, an astute observer will notice certain "tells." Their ears and tails give the game away, as do their eyes—the more white you can see, the more distressed the animal.

As with many other animals discussed in this book, the bond between mother and offspring is crucially important to cattle. In wild, or at least in feral, bovines, calves tend to be weaned at around six months of age, or still later in some cases. Even then, the mother and calf retain an affinity until the calf is at least a year old. In the dairy industry, it's common practice to remove calves from their mothers within a day of being born, and sometimes to raise them in individual pens. This isn't done out of wanton cruelty—keeping the calf with the mother means a loss of milk yield, so they're separated. Dairy farmers are pressured into cutting their costs at every turn by the ridiculously low prices they get for milk; when we pay more for bottled water than for milk, something's wrong.

The sundering of the cow and calf is not without consequences to each. If they're within hearing range, the mother and calf call to each other across the barriers. Stress hormones increase in the blood of both. Bereft of their calves, cows seem to divert their anxiety by rubbing themselves against objects. The calves, meanwhile, grow more slowly than they might otherwise. And there are other problems. Being raised in isolation means the calves struggle to form social bonds in later life, find it harder to cope with stress, and often become more reactive and aggressive, which is one reason why dairy bulls are so much more dangerous than their beef counterparts. If calves that are separated from their mothers are put together in a group, they tend to do much better. Nevertheless, they sometimes try to suckle from one another, as unsatisfactory substitutes for their mothers. That, in turn, can lead to problematic infections and abscesses. The practices of the intensive, modern dairy industry are an economic necessity for farmers, and the costs paid by the animals themselves are high.

There are possible solutions. One compromise between the need to maintain both milk yields and animal welfare is to allow contact between the cow and calf for at least part of the day for a while. But this doesn't come for free. At the heart of the matter is the question of whether you would be willing to pay more for your milk.

Most of us only see herds of cows from car windows as we drive by. In a snapshot like this, you could be forgiven for thinking there's little going on, just a bunch of large animals scattered across a field with an endless buffet of grass, enacting a monotonous cycle of walk, munch, walk, munch, enlivened only by not infrequent farts and other scatological eruptions. Domesticated farm animals are energy conservers, not given to unnecessary gamboling, and staid in their meanderings.

Lengthier inspection reveals a subtler truth. First and foremost, cattle are intrinsically social animals. Left to their own devices, they form herds that in structure are fairly typical of other grazing mammals. The core unit is made up of cows and calves, while mature bulls live either on their own or in small bachelor groups. On the farm, cattle tend to be kept in single-sex groups, and within those the animals develop affiliations and a hierarchy. Young calves hang out with others of their own age, while their mothers keep an eye on them, sometimes leaving the job to a babysitter, or guard cow, if there's a particularly succulent bit of clover in need of attention.

During their early weeks and months, the cohort of calves form relationships with each other and work out who's the boss. Once this is sorted, they can keep this pecking order pretty stable for the rest of their lives. They respect their elders—even mature cows that are smaller and lighter tend to boss the younger ones. In larger herds, cliques develop as groups of firm friends associate with one another. They don't necessarily roam far, partly because cattle are drawn to their birthplace, the part of their habitat where their mother reared them. So on large cattle ranches you might end up with local groups, drawn together both by geography and friendships.

Cattle are conservative—old cows really don't like surprises. They're creatures of routine and don't like novelty, whether that comes in the shape of new things in their environment, or new cows in their herd. Given a choice between known, familiar cows and a stranger, they pick the former every time. This makes it challenging to introduce new animals into an established herd as adults. The older cows get, the more stuck they are in their ways. It's a common trait in lots of animals, us included; the flexible approach of youngsters gets replaced by a dislike of change. You can see this yourself if ever you walk through a field. Mature cows tend to stay out of your way, but younger cows—heifers—will often crowd around you to investigate, seemingly daring one another to get closer and trying to explore you with their tongues.

As with other social animals, cattle are in tune with each other. They're quick to pick up on fear. An anxious cow doesn't even have to be on the scene to put the others on edge—the smell she leaves behind is enough. Counteracting this is exactly the kind of social buffering that acts like a comfort blanket to rats. Nervous cows seek out calmer ones for reassurance, especially familiar herd members with whom they have a close relationship. They groom one another, giving each other a tongue massage around the head and neck areas, and as they do so, their heart rates drop and they relax. The people who work closely with the cows become part of the herd's social environment, and the cows pick up on their emotions, too. It's a general rule that calm farmers have calm cows, while bad-tempered farmers end up with a skittish herd.

I think many people have an impression of farm animals similar to mine when I first moved to the countryside. They can identify with Neil deGrasse Tyson's tweet. To many, farm animals are part of the furniture, functional but not interesting, alive but not really sentient. We don't develop the close connection with them that we do with other domesticated animals—which makes it easier for us to eat them. For all that we think so little of them (if we even think of them at all), it's difficult to imagine the past ten thousand years of humanity without cattle, sheep, goats, and pigs. Our civilization has been built not only on our own sociality, but on the social animals we've brought with us on that journey. It would certainly be more civilized of us to recognize that they are far from stupid and that it behooves us, at the very least, to treat them with respect.

THE PACHYDERM PARADE

The great philosopher Forrest Gump once pointed out that "life is like a box of chocolates, you never know what you'll get." So it is when you're driving in the African bush. I was in a battered truck alongside a local guide when we crested a ridge to find a bull elephant in our path. The elephant stiffened and turned, then fanned his great ears out and walked toward us, a great gray edifice of indignation. Clearly, he wasn't accepting visitors. The walk turned into a loping, stiff-legged run, and the gap between us began to narrow alarmingly. While elephants often do mock charges, there's no such thing as being mock squashed, so the driver crunched the aged gearbox into reverse. Engine screaming, we shot backward at a pace that marginally but crucially outpaced the elephant. Satisfied that his message had been received, the elephant slowed, eventually coming to a stop, and then he began to take a dust bath in the middle of the road. He showed no inclination to let us pass, so, in our enforced wait—what the driver called an "African traffic jam"—we settled down to admire him from a distance.

Elephants are familiar animals to us, cultural emblems in many parts of the world, stars of the screen and zoo alike, but being close to one for the first time in the wild forces a reappraisal. The size and latent power of the bull was more vivid, more awe-inspiring, more thrilling here in his own realm than it ever could be behind the bars of captivity. This was his country. We waited at his behest as he showered himself with ocher soil, all the while issuing a rumbling bass commentary that seemed to reverberate in my bones. I lost all track of time, transfixed by the encounter and the indescribable privilege of being at close quarters with one of nature's masterpieces. Finally, appearing to catch a sniff of something in the breeze, he turned his head to the side and moved on, simply disappearing into the acacia scrub. He'd seemed huge, but judging by his tusks and his sprightly step, he was probably still a youngster. The fabled giant tuskers of old, who lost their lives to satisfy the burgeoning demand for ivory, could reach a colossal 10 tons, which means they were likely more than twice the size of the magnificent beast we'd just seen. How extraordinary they must have been, and how perverse that anyone could obliterate living marvels like these to make pointless trinkets. To my mind, it's the biological equivalent of destroying the *Mona Lisa* to get da Vinci's signature.

The root of all this is, of course, money. A few days after my encounter with the elephant, I had an even closer encounter with another of Africa's "Big Five." This term was originally coined to describe five of Africa's most charismatic mammals, the ones most sought after by trophy hunters: the elephant, the lion, the leopard, the buffalo, and the animal that I now tremulously stood next to, the rhino. This giant beast, second only in heft to the elephant among land mammals, was slumbering. I was invited to pat it and, cautiously, I did so. It felt more like a thing of granite than a living animal, its unyielding solid flank apparently oblivious to my touch. The rhino in question was a tame individual that had suffered the unkindest cut of all, losing his wedding tackle at a young age in a fight. Having been painstakingly nursed back to health, he was thought unfit to return to the wild, and now this behemoth lived under twenty-four-hour protection while the Kenyans worked out what to do with him.

The fact that armed guards were positioned at the entrance to his enclosure told its own story—the animal was worth substantially more dead than alive. In fact, this one rhino had a value of over three times the annual salary of the average person in that part of Kenya, all based on the ludicrous notion that its ground-up horn held a medicine that could, among other things, reawaken the potency of flaccid elderly men. The rhino's horn is, in reality, made mostly out of keratin, the protein that forms our toenails. Nevertheless, where there's demand, there's a market. The vast prices being paid for rhino horn at the time meant that by killing a single animal, a poacher could buy security for his family, a passport out of rural poverty. We can deplore the action, but we can perhaps understand the motive. That motive, fueled by insatiable demand, has seen the world's rhinoceroses reduced in number by 90 percent or more, making them locally extinct across large swaths of their former range. The same dismal story can be told of many other animals, including, of course, the elephant.

Although the commodity for which elephants are sought—ivory—is different, the pattern is the same. A lethal combination of greed, connivance, and ignorance put elephants in the firing line. At the peak of the market, the raw ivory of a single large elephant could be worth $100,000. Money talks: toward the end of the twentieth century, an average of two hundred elephants were being killed illegally every day. These numbers have declined in recent years, but that has more to do with a shortage of elephants in the poaching hotspots than with the enlightenment of fans of crass curios.

This isn't small-scale opportunism; it's industrialized killing on a huge scale. Poachers cross national boundaries, armed with high-caliber weapons and supported by an organized web of criminal infrastructure, to move bloody tusks along the supply chain. Progress in curbing the disgusting trade in ivory is slow, particularly in the Far East, though, following overdue legislation in China, prices have fallen dramatically in the past five years. As the rewards decline, the threat to elephants diminishes, but it hasn't gone away. And it won't until people realize that the true price of ivory is far greater than can be measured in currency—it's the persecution and loss of one of the most incredible animals on Earth.

Sadly, the market for ivory isn't the only threat to elephants. The spread of humanity puts a squeeze on the available land. The larger the animal, the greater the problem, because they need a larger range in order to thrive. So it's no surprise that elephants are on the front lines of this conflict. We know elephants are powerful animals; it's intuitive, obvious. Yet it's not until you see a fully grown tree that's been pulled to pieces by one that you get a real understanding of their strength: branches as thick as your leg smashed into splinters, smaller ones torn entire from the tree and eaten like so much broccoli, their vicious-looking thorns ineffectual. As I witnessed in central Kenya, browsing and grazing animals abounded, but few were quite so destructive as a posse of elephants. When they target farmland, the results can be disastrous. Growers can lose their livelihood more or less overnight, their crops trampled and eaten. A single elephant might eat a quarter of a ton of food in a single day, wiping out more than 2 acres of maize. Faced with this it's scarcely surprising that some farmers cultivate a strong dislike of elephants and a desire to protect their interests at whatever cost to the animals. As dangerously easy as it is for a Western, urban dweller to condemn, you have to put yourself in the farmers' shoes. It's a crucial issue because, for any conservation effort to work, the local people have to be on board. In this case, that means coming up with methods to counteract the world's largest cat burglars.

Other than by succumbing to the temptation to shoot, how might farmers dissuade a determined, massive animal with a taste for cultivated corn? A traditional option is to encourage hornets and bees to nest at the margins of their fields; this works to some extent, as elephants aren't fans of these insects. Electric fences offer another solution. Unfortunately, elephants in some parts have learned to negotiate these by dropping large branches on the fences, crushing them, and shorting out the power. There's even an account of one problem-solving elephant following the perimeter wires to their source and smashing the power base. Recently, some success has been had using elephants' dislike of chili. Briquettes made of elephant dung mixed with chili can be set alight. The resultant smoldering, spicy vapors are irritating enough to the elephants to turn them back without doing any lasting harm. More dramatically, you can pack a condom with chili powder and firecrackers. The explosions, accompanied by the fiery heat of the active ingredient of chili, capsaicin, provide an inventive deterrent. But I've also heard of ingenious efforts to use the elephants' social behavior against them. Oftentimes, the crop raiders are young males who have been evicted from their family groups, and who then seek to earn a living as best they can. These impressionable youths are sometimes led astray by following the bad practices of older, more seasoned burglar bulls. Yet in a scenario familiar to anyone with an affection for the writings of P. G. Wodehouse, these males can be browbeaten by the scoldings of an aunt, or indeed, any domineering matriarch. Playing recordings of the calls of such an adult female supposedly strikes fear into the hearts of these brave young robbers. I don't know how effective this is; more research is needed. Still, it's one creative approach among many designed to diffuse a conflict that, if unattended, could completely undermine attempts to protect elephants in their homelands.

Modern elephants were once simply divided into the African and the Asian varieties, but the study of genetics has provided us with a richer understanding. There's not one, but two African species—the savannah elephant and the smaller, less well-known forest elephant. The Asian elephant is intermediate in size to its African cousins and can itself be divided up into a number of subspecies. While I say cousins, Asian and African elephants have been distinct for millions of years—Asian elephants are more closely related to the extinct mammoths than to their living African counterparts. Aside from their extraordinary size, all share that most unique of all animal characteristics—the trunk, a conjoining of nose and upper lip into a dexterous, multifunctioning super-organ. Trunks combine the strength to hoist a baby elephant or a 600-pound tree trunk with the ability to pick up something as small and fragile as a tortilla chip without breaking it.

With a proboscis that big, it's perhaps not a shock to learn that elephants have a good sense of smell. In fact, it goes beyond merely good—elephants are stellar smellers. They have more genes specifically dedicated to this sense than any other animal, encoding an array of olfactory sensors that together provide them with an exquisite appreciation of the chemical world around them. This sensational sense allows the detection of standing water tens of miles distant, as well as the ability to make minute distinctions between food sources. In one test, elephants proved capable of determining which of two containers held the most sunflower seeds by smell alone. They have a nose for different peoples, too, and can distinguish between the smell of clothes worn by Maasai and those of Kamba ethnic groups. There's good reason for that, as young Maasai men sometimes spear elephants as a rite of passage; hence the animals respond to the smell of Maasai men with fear and aggression. Their sensitivity to odors is part of what makes them react so strongly to chili powder, of course, and it gives us other chances to deter them from coming into conflict with farmers. For instance, although the strategic placing of hives at farm boundaries is a widespread method of repelling the animals, the application of bee pheromones, the chemical cues of angry insects, at the edges of farms might do the trick more effectively.

Elephants live in societies with close and enduring relationships that have more in common with those of whales and apes than with other herding animals. At the heart of elephant society is the family, which most often means a group of anything from two to twenty related females and their offspring. The family members roam together, cooperating in the quest to find food and water, their close social bonds providing a source of support in hard times for young and old alike. First-time mothers have much to learn about bringing up a calf, so it often falls to the more experienced females to lend guidance. Although they spend most of their time together, there are times, especially during the dry season, when shortages of suitable browsing cause the elephants to separate temporarily and forage in smaller groups. When they reunite after any significant period apart, they play out an exuberant greeting ceremony, approaching one another at a run, bellowing out a cacophony of enthusiastic calls, and flapping their ears with the sheer excitement of it all. As an affirmation of close-knit ties in the animal kingdom, the elephant greeting ceremony is second to none, a vivid demonstration of the central importance that elephants place on their social relationships.

The expressive calling of elephants is part of a wide repertoire of vocalizations. Though we're familiar with the squeals, trumpetings, and rumblings, there's a whole lot more happening when elephants communicate that was until recently a mystery to us. Experienced elephant watchers noticed that even when they're spread widely across the landscape elephants seemed unusually responsive to one another, and that small herds of them might occasionally stop in mid-stride and linger a while in apparent silence. Something was going on, but what? Were they conversing telepathically? The riddle was solved in the 1980s when a team of researchers led by Katharine Payne of Cornell University discovered that they were using infrasound—bass tones so low in pitch that they pass below the range of our hearing. Elephants that seemed silent to us were in fact engaged in clandestine chatter. Infrasound is an excellent way of keeping in touch remotely. Deep sounds have long wavelengths, which allow them to transmit over massive ranges and provide a fantastic means of making long-distance contact calls when families are scattered. In the right conditions, elephants can detect one another even when separated by a distance of 6 miles, and may be able to communicate more detailed information over 1 or 2 miles. The elephants' infrasonic sounds travel not only through the air, but also under the ground, as seismic waves. To listen to them, they stand stock-still, using their gargantuan feet, and sometimes their sensitive trunks, pressed against the dusty terrain. The sound waves are transmitted through their bodies, via their skeletons, to their ears. In this way, elephants collect and pass information among their dearest, if not necessarily their nearest. Encoded within the messages is a wide range of information: they learn the whereabouts and movements of other elephants and even the specific identity of who's calling. They're able to recognize perhaps as many as a hundred other elephants based solely on the contact calls. It's an extraordinary thought that conversations between elephants take place across landscapes without us even being aware of it.

Elephant families are built around a single leader: the matriarch, an older female with a lifetime of experience. Ascending to this status isn't a matter of competition or assertiveness; rather, it comes through the confidence invested in her by the rest of the family. Elephants are long-lived animals; their allotted time on Earth matches ours. For most of this time, females associate with the same individuals, an inner circle of close relations who know each other intimately. Thus the character of the matriarch is well known to the rest of her family, and this is crucially important—they choose to follow her for the enduring qualities of knowledge and good judgment that she brings to her leadership. In the punishing dry season, the quest to find fresh forage and a reliable source of water becomes a matter of life and death, particularly for the calves. All the adults in the herd contribute, but ultimately the greatest burden rests with the matriarch. She possesses a vast store of information, comprising that which she has learned and accumulated throughout her years alongside a legacy of wisdom from previous generations, resulting in a lasting elephant culture. Under the matriarch's guidance, the family traverses huge distances between favored areas of their habitat, revisiting particular clearings and reliable watering holes. In doing so, they follow long-standing, traditional routes that have been passed down from their ancestors and committed to memory.

Similarly, when danger threatens the youngest calves, veteran matriarchs prove the most adept at judging the risks. Large and powerful male lions, for instance, present a greater menace to the family's youngsters than do female lions. The matriarch understands the risk all too well. If she senses the presence of male lions, she brings her kin into close formation, the youngest hidden behind a protective cordon of adults, and guides them to safety. Given the longevity of elephants, a senior matriarch might see two generations flourish under her. Having Grandma in the family makes a staggering difference to the calves—they're up to eight times more likely to make it to adulthood if she's around.

As the chief influencer in the family, the matriarch plays a huge role in deciding where the group will move and when. She instigates a group movement by facing in her chosen direction and making a deep, rumbling "Let's go" call. If the group fails to keep together, she'll wait for them to catch up, peering backward at them and encouraging laggards with further rumbles. She isn't the only one to do this—elephant society is about support rather than tyranny—though she is the most active in marshaling them. The matriarch is a central, stabilizing figure, the glue that binds the family together.

That's never more clear than when, inevitably, the old lady dies. Leadership doesn't automatically pass to the next in line by age. A choice must be made, and only a well-respected elephant can successfully transition into the position of matriarch and unite the family behind her. It doesn't always happen smoothly. Bereft of their figurehead, elephant families sometimes split into different factions that each go their own way. Sadly, the imposing size of the matriarch often makes her a target for poachers. Decades of accumulated wisdom can be lost in a moment to a single bullet, casting the remaining family into a wilderness of confusion.

The spectacular sight of savannah elephants gathering in their hundreds is rarer than it once was, but families periodically converge, and when they do it's an opportunity for more distantly related elephants to renew old relationships. This is part of what's known as a fission-fusion society, the merging and splitting of animals within a population into groups and subgroups. Though the family is the day-to-day building block of elephant society, these clan gatherings share genes and culture.

The web of relationships among a population of elephants thus extends far beyond the family unit to encompass individuals over a vast geographic range. As the super-herd dissipates and the various families go their own way, the elephants occasionally switch groups. A young male on the cusp of independence might eschew his mother's company to tag along with someone new. Or a female may decide to link up with her cousins in a different family, especially if her ties with the females in her own group have weakened over time. As they disperse over large distances, beyond the limits of their deep-sound communication, elephants may lose contact with each other for extended periods. Gone they may be, forgotten they most certainly are not. The elephants' prodigious and legendary memories see to that. The calls and the smells of long-separated elephants induce powerful and touching responses, even decades after the individuals last met. In one case, playbacks of the call of a female elephant who had left her family unit twelve years earlier were greeted noisily by her former associates. In another, the smell of their mother produced a stirring, unmistakable response in adult zoo elephants nearly thirty years after they'd been parted from her.

As youngsters, male elephants live alongside their sisters under the protection of their mother and the matriarch in the family. However, as they approach their teenage years, they begin to spend increasing amounts of time away from home before finally becoming fully independent and going solo in their teens. It's a tough time at first for the males, outside the supportive environment of the family. Imposingly large though they are, younger bull elephants face dangers on their own. In Botswana's Chobe National Park, there's at least one pride of lions who specialize in taking on these gargantuan prey, swamping their target in a concerted and coordinated effort. The risk to the lions is enormous, yet it is matched by the payoff should they succeed.

Adult bull elephants are sometimes painted as loners who leave socializing to the females. Closer observation reveals that they do seek out company and that, just like the females, they tend to associate most closely with relatives of the same sex, especially those of a similar age. Younger males spar with each other, testing their prowess for later in life, when fighting for dominance becomes of supreme importance. When the shoving is done, they will go shoulder to shoulder in coalitions to stave off intimidation from outsiders. Within these bachelor clubs, older males play a crucial role, in some ways like the older matriarchs of the family units. These gigantic tuskers can be surprisingly playful—huge bulls have been known to get down on their knees to spar with younger males. It's not all high spirits within male groups, though. In the dry season, especially, when conditions are at their harshest, there's a strictly enforced hierarchy, which is most obvious when you watch younger males take their place in the queue for access to water. They greet the dominant male with an elephant's version of a salute, placing the tip of their trunk in his mouth as a gesture of subservience. Yet even with this hierarchical system, when they're at the watering hole and have paid their dues, the males are surprisingly tactile, draping their trunks over one another's backs and flapping with their giant ears together in a kind of elephant high-five. It's noticeably different from the behavior of the female-led family groups at watering holes. Perhaps it's because of the need to remain wary of approaching threats to their calves, but the females are much more restrained when they're down for a drink.

Despite the easy familiarity of interactions between younger bulls, male elephants undergo a periodic Jekyll and Hyde transition as they age and mature. Even the most peaceable bull becomes a maddened and dangerous beast when periodically he enters a state known as musth, supercharged by a flood of testosterone through his system. Anything that gets in his way may fall victim to his capricious rage, and there are stories of bulls in this state going on killing sprees and slaughtering dozens of rhinos. Hormones are a major part of the aggression surge, though it's also been speculated that the swelling of glands in their temples at this time compresses their facial nerves, causing such intense pain that it drives the hyped-up males into a kind of madness.

A male in musth is hard to miss. A sticky substance oozes from just behind the eyes, giving the male a dark-cheeked appearance and a characteristic filthy stench that makes a teenager's bedroom seem like a perfumed salon by comparison. If those in the vicinity needed any further convincing to avoid a male in a musth funk, the elephant provides it by dribbling a trail of equally foul-smelling urine as he walks. In this aroused state, males are prepared to take on even those bulls to whom they've previously been subservient, shattering the relative harmony of usual elephant life.

A pair of fully grown bull elephants fighting is a terrifying spectacle of brute force. It's triggered by competition for mates. Females coming into estrus produce an alluring smell accompanied by a coquettish rumbling call that excites nearby bulls into action. Prospective paramours converge, and that's where the trouble begins. Arriving at the scene, a bull finds another already in residence, and in their desperate desire to make little elephants, violence flares. The rivals confront one another, each attempting to intimidate the other by kicking great clouds of dust in the air, issuing bellowing challenges, and flaring his ears to emphasize his bulk and might. If they're well matched in size, then the two behemoths go head to head in a titanic shoving match, a combined 12 tons of muscle and aggression concentrated at the point where their bodies meet. Tusks clash like thunder cracks as each bull tries to assert himself and gain advantage. Close contests may be drawn out, pushing both to the edge of exhaustion. Finally, the resolve of one combatant breaks. He displays his submission and turns tail. Normally this is the end of the matter, but when the victor is in musth, he's an enemy to reason, and may refuse to accept the surrender. When this happens, he may pursue his opponent with murderous intent, gouging him with his tusks to inflict severe, sometimes lethal injuries.

As a counterpoint to this violence, the nurturing, protective behavior of these intensely social animals in their normal state is compelling. Huge though elephants are, their inexperienced calves are vulnerable and need help to negotiate life's challenges. This is where a little know-how within the family comes in especially handy. Though the mother is the most important figure in a calf's world, the rest of the family, and particularly the most experienced females—perhaps aunts or great-aunts—are supportive and sensitive. Rearing calves is a team effort and is central to family life. They babysit calves, allow the youngest to suckle from them on occasion, and retrieve calves that have wandered too far from the herd. Elephant helicopter mothers even step in to prevent excessive roughhousing among boisterous calves. Intriguingly, these vastly intelligent animals don't simply react to a calf's cry of distress; often they anticipate trouble before it happens. They seem to have an appreciation of what might constitute danger to smaller family members, rerouting their trips to avoid swampy ground or deep water, in order to account for the differences in their capabilities. And if a youngster does get stuck in cloying mud or at the foot of a steep riverbank, they may hoist it to safety with their trunks, or rearrange the bank with their tusks, to make a shallower gradient for little legs to manage.

Assistance isn't solely offered to the young—the adults support each other, too. We most often see this in the context of our interactions with them. In some populations, it can be a challenge to tranquilize elephants effectively, because if an elephant sees a dart embedded in her kin, she'll most likely pluck it out. Darting elephants, however, is a necessary part of managing elephant populations in the modern world. Constraints on the movements and migrations of elephants in protected areas mean that sometimes we have to intervene to translocate animals and so preserve the genetic diversity of populations.

The strength of elephant social relationships can make this a challenge. A famous account from the Anglo-Dutch wildlife veterinarian Toni Harthoorn, of an elephant family's reaction to one of their number being darted, paints a typical picture. He managed to hit the target with a long-range shot from the dart gun, and when the tranquilizer took hold, the animal, as expected, slumped to the ground. Most animals, already spooked by the presence of a shooter and the fall of a companion, would rush off in a panic. Not elephants. Harthoorn described how the family responded to his intervention with an "indescribable melee of screaming" while trying to hoist their stricken relative to her feet. The powerful drug he had used was hundreds of times stronger than would be needed to kill a human, and it put the elephant out of action for two hours. Her family kept trying to lift her for the entire time. Eventually, when she was able groggily to support her own weight, they flanked her and took her to the safety of a nearby stand of trees. A united front of elephants makes getting access to a single animal an immense problem: the protective instinct of elephants toward their own is unwavering.

Interventions on behalf of injured or imperiled family members show the close affinities that exist between elephants. There are even accounts of them helping out unrelated elephants in the same way. They don't draw the line at helping their own species, either. There's an anecdote of a working elephant refusing to lower a heavy wooden pillar into its mooring because, at the foot of the hole, a dog lay sleeping. Remarkably, given our far-from-perfect relationship with elephants, they've even been known to intercede to help people. Five years ago in West Bengal, an elephant entered a village and, perhaps in anger, perhaps in confusion, destroyed a house. The members of the family within were unhurt, but a baby in a cot, who'd been in the part of the building affected the most, was trapped. The baby's cries brought the retreating elephant back to the house, whereupon he carefully cleared away the rubble, allowing the family to recover the infant, who was unaffected by the episode. The elephant apparently recognized a cry of distress when he heard one. And it seems that crying itself isn't a uniquely human behavior—it's one that has played a part in numerous elephant stories.

The most touching of these tells of a newborn elephant who, in 2013, was rejected by his mother at a park in China. In fact, not only did she reject him, it seemed as though she wanted to harm him. As a precaution, the keepers stepped in and separated the pair. Although it did not suffer lasting injuries, the baby elephant was so traumatized by the experience that he cried nonstop for five hours. He presented a pitiful image, lying under a blanket with tears rolling from his eyes.

What are we to make of these kinds of stories? It's a very human response to see direct parallels between the baby elephant's apparent grief and our own experience, or to conclude that the bulldozer elephant had empathy for the human baby that he'd inadvertently imperiled. But at this point, the question of whether elephants possess emotions similar to ours is one for philosophy rather than science. A scientific evaluation would require us to amass evidence under carefully controlled conditions, which are plainly lacking in these anecdotes. However we might want it to be otherwise, we can't conclude with certainty that what underlies these behaviors is a human-like range of emotions. Then again, nor can we state definitively that elephants don't possess these finer feelings. What we can say is that elephants, in common with a growing number of other animals, have a previously underappreciated depth both in their cognitive abilities and in their emotional lives.

We see the most extraordinary side to elephants in the solemnity with which they greet death. Although the elephant graveyard of legend is no more than a myth, elephants clearly respond to the bones of their own kind. There are even reports of them showing a disquieting curiosity toward ivory, reaching out to inspect trinkets made from their dead. But the most poignant displays of emotion occur in the context of the death of a family member. An elephant funeral provides the clearest picture of an animal that shares with us a sense of loss. A ranger in the Serengeti described one such ceremony to me, recalling how an elderly matriarch was manifestly failing in the endless weeks of a punishing drought. As the vegetation withered daily and the wind whipped the soil to dust, it was clear that this was to be her last dry season. Gaunt and shambling in her weakness, the slow march of her final days saw her family of four adult females, most likely her daughters, lingering at her side. On the last day she was seen, she could barely move at all, and sometime that night, she fell for the last time and died.

Dawn brought the sight of her family gathered around her, eerily quiet and somber, caressing her body and observing a ritual that seems to be widespread among elephants. After standing in silent vigil for several hours, they began to cover her body with branches, leaves, and earth. They remained as the shadows grew longer and another night approached, moving very little, except, in one case, to scare off an inquisitive jackal. The following night they moved on, no doubt driven to find the forage and water they'd had to forgo while attending their struggling matriarch. Some weeks later they returned to the site, and though little remained by then of their fallen relative, again they displayed the same quiet, respectful attitude. Though we can't know what goes on inside the mind of another animal, the elephants seemed to show the hallmarks of grieving, and in a way that we can immediately recognize. It suggests that these extraordinary animals have an awareness of mortality, and therefore, possibly, a conscious understanding of what it means to be alive.

Not so long ago, elephants ranged across huge areas of our planet. They covered the vast majority of Africa and lived across Asia from Iraq to China, as well as down into Indonesia and Borneo. Just two hundred years ago, there were an estimated twenty to thirty million elephants. Now there are, at most, half a million left. At the same time, human populations have mushroomed. There are twice as many people in Africa today as there were forty years ago. Along with this rise in population has come an ever-increasing demand for space, not only for farming, but for infrastructure, such as roads that bisect the landscape. As the remaining elephant populations have dwindled, so, too, their range has contracted, from continental scales to an archipelago of isolated pockets scattered within the great swath of the lands they once roamed. These islands of elephant habitat are being cut off from one another by the rising tide of human development, so that elephants are now ever more hemmed in. This isn't a problem solely for those who live alongside elephants—it's for all humanity to solve. If elephants are to have a future, it's going to take a concerted international effort to address the problem. It would be a horrific indictment of our species if, just as we uncover the hidden complexities of elephants, we were to wipe them off the face of the Earth.

CHAPTER 7

BLOOD'S THICKER THAN WATER

Cooperation among relatives is often the
secret to success for carnivore clans.

A ROARING SUCCESS

It's my first night in Africa and I'm lying in a banda, a traditional circular hut, listening to animals who, like me, aren't sleeping. I'm on red alert. My fellow occupants of the banda, two other researchers, arrived before me and very wisely chose the beds away from the window. My head is right next to the window, which is unglazed. It does have bars on it, but that doesn't reassure me. The bars won't stop an adventurous baboon from reaching through to tousle my hair; nor will they stop an enterprising leopard from reaching through to reconfigure my face. The generator has been switched off for the night and all is darkness. The only sounds are those that drift in through the window. Here, in the central Kenyan bush, those sounds are all made by animals. There are hoots, cackles, grunts, and yaps alongside the rustlings of vegetation and breaking twigs. These are all new to me and entirely unfamiliar. I decide that the cackles belong to hyenas, which is exciting because I'm desperate to see these misunderstood and fascinating animals.

And then I hear another sound—one that needs absolutely no introduction. It's primeval, stirring feelings deep inside me. Even though it's distant, the sound of a lion's roar in the darkness is stunning, unique, terrifying. It transports my mind instantly to a book I read describing the man-eating lions of Tsavo, not a million miles from where I lie. Over a period of several months in 1898, a pair of lions terrorized a camp of workers building a railway bridge over the river Tsavo. Under the cover of darkness, the lions would steal into the camp to seize and carry off their screaming victims from their tents. Something like thirty workers were taken to their deaths until the lions' reign of terror was ended by John Patterson, a British army officer in charge of the project, who kept a nightly vigil from a platform by the camp. The lions were tenacious, though—one of them was shot six times before it succumbed. With this thought in my head, there's no way I'm going out to the toilet block tonight.

It's tempting to ascribe my visceral response to the lion as some echo of my ancestors. There was once a time when lions could be found not only throughout Africa, but also in southern Europe and across the Middle East, and into Asia as far as India. Its extinct relative, the cave lion, was widespread throughout Europe; their remains were even found during excavations under Trafalgar Square in the 1950s. Many generations ago, my forebears lived alongside these animals, and they didn't have bars on their windows. While there are relatively few animals who might regard humans as prey, lions are potentially one such. Perhaps it's not surprising that I reacted to the lion's roar after all. Despite this, I can't help feeling a sense of sadness that the modern distribution of lions is vastly reduced from what it once was. Lions can now only be found in pockets of sub-Saharan Africa, plus a small remnant population in India. Excellent predators though they undoubtedly are, they have no effective answer to modern humanity's weapons of wildlife destruction.

Eventually, despite the adrenaline surge induced by the lion's roar, sleep claims me, and I wake up unmolested by the local wildlife, with a relatively normal face. Along with my two companions from the banda, I'm heading off on a dawn game drive, going out into the Laikipia bushlands in trucks to spot animals. After an hour or so, we stop on some high ground. My head is buzzing with the diversity of creatures I've seen in the past hour. I feel elated—it's like nothing I've ever experienced before. A hartebeest nurturing her newborn calf, nudging it to its feet. Vervet monkeys bouncing around the trees like hyperactive children. The surreal sight of three giraffes running, covering great distances yet seeming to move in slow motion. All of it is new and overwhelming. Nowhere on Earth can compare with this part of Africa for its breathtaking mammal megafauna. Seeing it all from a truck is one thing, though. Stepping out into the bush means shedding the protection—real or imagined—provided by the truck. Still, we have experienced local guides with us, and, of course, they'll have guns, won't they? I decide to ask. No, they're not carrying guns. One of them has a pointy stick. At this point, I call to mind the signs that zoos put up: "Don't feed the animals." It takes on a different complexion here.

I have a decision to make. I can stay on the truck and be sure of not feeding the animals, or I can go into the bush. At this point, the biologist in me comes to the fore—I want to experience this, even if the risk isn't quite what I'm used to in my closeted modern life. Besides, the guides are entirely unconcerned. They've grown up here, and their practiced eyes read the landscape like I'd read a book. Reassured, I wander off, and as I reach a thicket of vegetation, there's a sudden explosion of activity immediately ahead of me. Three large animals burst from cover and charge toward me. I don't drop into a protective crouch, or spring for cover. Instead, the chief aspect of my instinctive defense is to freeze in abject terror even as they rush past me. If this had been a trio of lions, it wouldn't have required their famed cooperation to capture me. As prey, I would have been about as challenging as a pork pie. To my relief, though, the animals aren't interested in me, or pork pies; they are bushbucks, a kind of medium-sized antelope, and committed vegetarians. Still, no more wandering off on my own.

Over the following days, I developed into a more seasoned bush-walker: still hopeless in comparison to the local guides, but at least not so prone to doing stupid things. Besides, talking to the guides forced me to reappraise the risks posed by different animals. Despite the plethora of potentially dangerous ones, the consensus was that they all represented a manageable risk. There were, in fact, only two mammals in that part of Kenya that caused serious consternation—leopards, phenomenal predators by night, and buffalo, which have been known to ambush and gore the unwary. The locals certainly had a deep respect for lions, but that respect didn't translate into fear. In the guides' long experience, so long as you gave them a wide berth, they presented little danger, at least during the daytime.

Humanity isn't always quite so enlightened when it comes to li-ons, however. For trophy hunters, they are a kind of gold standard. People come thousands of miles and pay tens of thousands of dollars to courageously shoot them with high-powered rifles from massive distances. Bravo. At the same time, they have been ritually hunted as a rite of passage by some of the peoples living alongside them. Mean-while, the decline of the tiger has encouraged the exploitation of lions to satisfy the demand for the entirely imaginary medicinal qualities of big cats that are peddled as alternative remedies in some parts.

Yet the greatest cause of decline is the changing face of Africa. The number of people in Kenya and neighboring Tanzania has increased something like tenfold over the past hundred years. Expanding hu-man populations have intensified the pressure on land resources, leading to habitat loss and a decrease in the numbers of prey. As ever more people have chosen to live in close proximity to lions, the rela-tionship between our two species has become increasingly fraught, sometimes in unexpected ways. Distemper passed from domestic dogs killed 30 percent of the lions in the Serengeti National Park in 1994. Grasslands that once supported wild mammals have gone under the plow. Faced with fewer of their natural prey to support themselves, lions have escalated their attacks on both livestock and people. In response, farmers poison the lions.

The decline has been staggeringly rapid. From around half a million wild lions in the middle of the twentieth century, there are perhaps only twenty thousand today. Alongside this implosion, populations are fractured by structures such as fences and roads. Lions, like many other mammals, disperse to find mates. If this is no longer possible, the result is inbreeding and a loss of genetic diversity, which in turn increases the animals' susceptibility to disease and reduces their fertility. The lion's tale of woe is one of loss and destruction. Many deeply committed people have taken up the challenge of addressing the lion's plight, but it remains to be seen if these efforts will be enough to save this most emblematic of species from extinction in the wild.

In human culture, lions are a symbol of power and courage, strength and chivalry. Despite, or perhaps even because of, their occasional tendency to snaffle people, they are depicted almost exclusively in a positive light, particularly throughout their original range in Africa, Asia, and Europe. Singh (which means "lion") is the sixth most common surname in the world. Thirteen popes were named Leo. Singapore ("lion city") is named for them. The list is endless. Partly, this relates to some impressive statistics. Fully grown, a male lion is an imposing beast—more than 400 pounds of power, standing around 4 feet tall at the shoulder. While females lack this bulk, they're more agile than the males and are powerful animals in their own right. The massive head and jaws of a lion yield a bite that is something like five times as powerful as our own. Four wickedly sharp canine teeth as long as your finger can puncture tough hide. Toward the back of the mouth, carnassial teeth can slice like guillotines through skin, tendons, and bone. Paws the size of a plate, fringed with viciously sharp claws, complete the lion's armory. All of this combined with incredible acceleration, and a top speed around twice that of an elite human sprinter, make lions among the most formidable hunters on Earth.

The statistics are only part of the story, however. Another major element of the lion's success, the reason they first spread from South Africa to Greece and from Senegal to India, is group work—they are the only truly social cat species.

Why, among all the big cats, are lions alone social? Plenty of reasons have been suggested, but the one dominant factor is, ultimately, other lions. In the turf wars between lions, there is strength in numbers. The best real estate for lions is a waterfront property, a place that allows the lions access to water and to shade under the lush vegetation that tends to grow in such places. Just as importantly, living near water means they control an area where their prey must come to drink. A solitary lion is unlikely to be able to hold on to such a prime spot against a pride. Prides beat singletons, and large prides beat small prides. By banding together, lions gain bargaining power for the best territories.

Winning a territory, however, is only the first step; it then must be defended. The resident lions in a pride advertise their possession by scent-marking and roaring. The roar is astonishingly, painfully loud. At 114 decibels, it's about as loud as the siren on a police car, and it carries a potent warning for up to 6 miles across open country. Scent-marking is scarcely less dramatic. Pride males spray a prodigious and pungent mix of urine and pheromones onto the bushes, trees, and rocks that ring their territory. The freshness (or perhaps *recency* is a more appropriate word here) of the smell backs up the challenge provided by the roars. As if that weren't enough, the lions also mark their territories by scratching and gouging marks on prominent vegetation. The effort that goes into declaring ownership serves to ensure that any passing lions get the message: keep out, trespassers will be macerated. Other lions in the area may be persuaded to leave well enough alone. Or not. Rival lions gauge the size of the home team by the roars and signals they provide, and this tells them their chances of success if they challenge that pride for their territory.

There's much more to lion sociality and cooperation than territorial defense, though, and to uncover this, we need to delve into the amazing world of the pride.

Perhaps the main difference between mammal groups and other social animals, such as birds or fish, is that mammal groups are usually formed around a nucleus of adult females. Lions are no exception. A pride of lions has at its heart a number of lionesses. Often these are closely related. Several generations of lions can be found in the same pride—daughters, mothers, grandmothers, and even great-grandmothers in some rare cases. The adult females aggressively exclude other females from joining, so a pride remains a family affair. Their close relationships can occasionally extend to nursing one another's cubs, even adopting orphans whose mothers have been killed during a hunt or a confrontation with another pride.

For such a deadly, ruthless predator, the strength of the mothering instinct seems somehow paradoxical. In a few rare instances, female lions have adopted a calf of one of their prey species. Days-old oryx, springbok, and gazelles have each been seen in the presence of a female lion, who tends and protects them as though they were their own cubs. Often this kind of adoption happens after the lioness's own cubs are killed, which might give a clue to her mental state. The pioneering behaviorist Konrad Lorenz noted years ago that certain characteristics of young animals tend to provoke a nurturing response. Features such as a disproportionately large head, small snouts, and large eyes tend to activate this response—it's one reason why teddy bears, for example, look the way they do nowadays: Teddy bears in the Victorian era were more bear-like than their more modern, cutesy counterparts. Perhaps packing the kids off to bed with a realistic copy of an apex predator was all part of toughening them up back in the day. As time went by, toy manufacturers realized that children were quicker to form attachments to cuddly toys with an infantile appearance, and also more likely to pester their parents into buying them. So the teddy bear morphed over time until it reached its current, unthreatening appearance. We're preprogrammed to go easy on baby animals, and this might also explain why lions occasionally adopt calves. For all that, a lion pride doesn't necessarily provide the best childcare environment for a baby gazelle. The protective instinct of one female can only go so far when the other pride members are feeling peckish.

Despite their all-action image, lions spend an awful lot of their time doing very little. For around twenty hours of each day, they may be found relaxing under the shade of a tree. The lion siesta site can be full of conversation. Lions are highly vocal animals with a range of different calls. Grunts, groans, whines, snorts, howls, moans, and even meows are used in different contexts to signify the feelings of the speaker. Amid the chattering, other social activities are taking place that reinforce the pride's bonds, such as nuzzling and licking one another's heads. These mutual grooming sessions occur mostly between individuals of the same sex, so females groom females, and males groom males. This apparently simple pattern reveals to us the truth of a lion pride. We might think of it as a coherent team with strong, mutual ties. The reality is that the pride is more of a loose alliance between two tightly bonded groups, one of closely related females and the other a smaller group of males, unrelated to the females but often related to one another. Though males will come and go, the lionesses are a constant in the pride. Unusually among social mammals, there's little or no dominance hierarchy among these female lions. All have equal status within the pride, and each produces a similar number of cubs. The males are not so egalitarian: there's usually a dominant male, and he gets to father the most cubs.

The pride, then, is a mix of adult females, their cubs, and the dominant males: often around a dozen lions in total. Female cubs will assume their own position within the pride as they grow to adulthood, but male cubs face a very different future. As they approach maturity at the age of around two years old, they leave—or are banished. Their coming of age is signaled by the beginnings of a mane, and it's suggested that this might be one of the triggers to drive them out. In the face of this new, uncertain world, the evictees—brothers, cousins, and other males of the same age—band together for support. There might be just a pair of them, or as many as seven. Regardless, for the next few years, perhaps for the rest of their lives, their reliance on one another will be total. As they fill out, so their manes become full and lustrous, the Samson-like symbol of male lion potency. It was long thought that the mane served to protect the neck and shoulders from injury in the many battles that lie ahead in life, but there is little evidence to support this. Instead, it seems that the mane is a signal to other lions. A dark, lustrous mane is a feature of a lion with plenty of strength and vigor. It tells other males that this is a dangerous rival, and it tells lionesses that he's appealingly, dangerously sexy.

The coalition of young males live a nomadic, bachelor existence in the social wilderness. Over time, they develop the skills and strength needed to claim their own pride by ousting the resident males, and for that, the nomads must be in the condition of their lives. Timing is everything. Too young, and they lack the power to drive out resident males; delay, and before long they themselves will be vulnerable to the challenge of a new generation. The nomads eavesdrop on nearby prides, taking their time, calculating their chances. Selecting the right pride is the biggest, most important decision they will ever make. If they get it right, they can secure their bloodline. But if they underestimate their opponents and lose their bid for control, they could pay for the mistake with their lives.

As they cross the pride boundary, the nomads keep together in tight formation. They're cautious, yet brimming with youthful vitality and fearsome intent. They are heading for the epicenter of pride territory. The sight of the nomads stirs the resident pride males to action. They meet the trespassers with powerful, shuddering roars. The roars tell the nomads that they will not be tolerated, that the pride will be defended. This is the last chance to withdraw; failure to do so means the situation will in all likelihood escalate to lethal violence.

There's an electrifying cacophony of rumbling roars as the first skir-mishes break out. Rising up on their haunches, the rival males trade savage blows. Deep, raking wounds well with blood. In the intensity of battle, each side attempts to deal decisive, crippling injuries on the other. The nomads begin to get the upper hand. Now, as the con-test begins to slide away from them, the resident males face a new problem—how to escape in one piece. The victors press their advan-tage home and the clash turns into a rout. The dethroned pride males must retreat, fast, or else they may well be killed. If they make it out alive, they may regroup for a future campaign. Mature male lions bear the marks of previous encounters, dark furrowed scars across their faces and flanks. With so much at stake, contests for the su-premacy of a pride are bloody affairs. And the violence isn't over yet.

It's reckoned that males have only two or three years at their peak, the time during which they can hold the pride. There are parallels with this in the boxing ring, where the average reign of a male heavy-weight champion is around two and a half years. It reflects the simple fact that there's only a brief period of maximum physical prowess. For male lions, the stakes are higher than for boxers: they can only breed while they retain their control of the pride. Newcomers they may be, but the clock is already ticking. Female lions, like many other mam-mals, don't ovulate while they are producing milk for their young. The males can't afford to wait to let nature take its course, so they must force the issue. Shockingly, from a human perspective, they achieve this by killing the cubs.

Lion infanticide isn't a manifestation of wanton cruelty; it simply represents the harsh reality of life for male lions. Though the females vigorously defend their cubs, the greater size of the males places a limit on what they can do. Larger cubs may escape the carnage by fleeing, though the prospects are bleak for those who are still nursing. It may seem extraordinary to us that the adult females consort with the very same males that have killed their cubs, but it's a fact of lion society. Soon after the loss of their offspring, the females come into estrus and the race begins to raise another generation.

If the new resident males can maintain their hold on the pride, the next generation of cubs has a chance of making it to adulthood. Even in a stable pride, however, they do face danger. Wild dogs, hyenas, and leopards will all kill unprotected cubs, given a chance. Part of a mother lion's role is to move her young around frequently between different den sites, carrying them by the scruff of their necks, to avoid a buildup of scent that might allow predators to home in on the youngsters. Despite their reputation as the "king of the beasts," only around one in five cubs makes it to adulthood.

In the open grasslands of the Serengeti, gazelles, impala, wildebeests, and zebras remain alert as they graze. Well they might, with lions ready to strike at any time. Nonetheless, these are no easy prey. The many eyes of a herd are excellent at spotting approaching danger, and these long-legged herbivores have an impressive turn of speed. Lions can counteract their defenses by hunting at night: darkness makes lion-spotting harder, and the cooler conditions can make the prey more sluggish. Even then, the result is far from certain. Depending on the season and the prey, the likelihood of a kill is no better than one in three. Not only might they fail to bring down their prey, but a horn or a flailing hoof can damage a hunter. A broken leg, serious puncture wound, or fractured skull in this unforgiving country can mean a long, drawn-out death. But in the never-ending struggle between predator and prey, lions have a trump card: cooperative hunting.

Although lions can and will attack on their own if they sense an opportunity, their hunting prowess is greatest when they make a concerted effort. The likelihood of a kill increases as the involvement of pride females does, especially when the group is made up of experienced hunters. In Botswana, lions use this team ethic to spectacular effect, combining to take on and kill elephants, animals that are up to fifteen times heavier than themselves. Although lions are born with the instinct to stalk and hunt prey, they need to learn the skills that will make them effective killers. Mother lions sometimes present their cubs with an easy target, such as a calf, that the cubs can chase and harry. If the calf escapes the inexpert attentions of the cubs, their mother will retrieve it for another go. This pitiless exhibition can be tough to watch, since the victim is toyed with over an extended time, but the cubs have to learn. Eventually, the youngsters must graduate to the real thing—taking part in a real hunt. While mistakes made by the apprentices can cost the pride the chance of making a kill, the experience will serve the youngsters well in the future, so this is short-term pain for long-term gain.

The success of the hunt is down to teamwork and coordination. In Etosha National Park in Namibia, lionesses have honed their strategy to an art form. Approaching a herd of their prey, the pride members use the cover of vegetation to maintain the element of surprise. One or two lions move stealthily to each side of their target, bellies close to the ground, a total focus on their quarry. As these wings form a pincer around the animals they are stalking, the rest of the females remain hidden in the center, ready and alert for the attack. Each lion has her own favorite position as either a center or a wing, and the hunt is most effective when they all take their own specialist places. With pains-taking care, the wings edge forward. Every extra foot they can gain before they are discovered or break cover is priceless. Finally, all the lions are in place and the trap is set. Suddenly, they launch the attack. The lions on the wings charge, and the herd scatters in panic and confusion. Some of the prey are inevitably pushed toward the center of the lion formation, where a hidden welcoming committee awaits. If luck is on the side of the lions, they can ambush a terrified animal as it runs into them. An experienced lion can rapidly make the kill, crushing the unfortunate victim's windpipe, or, for larger prey, clamping its mouth over the prey animal's mouth and nose to suffocate it.

Pride relationships at the kill can be highly charged. Males, by dint of their size and rank, demand—and get—first access, regardless of their involvement in the kill or lack thereof. To add insult to injury, the dominant males may monopolize the carcass for hours, keeping the females and cubs at bay by threats and brute aggression. The low place of youngsters in the lion hierarchy can lead to starvation in harsh times. Having males at the kill does have one upside for the rest of the pride—they are far less likely to lose their prize to competitors such as hyenas, who are attracted by the commotion.

Although it is sometimes assumed that the females shoulder the burden of hunting in a pride, this is wide of the mark. Males do hunt; they just do so in a different way. Effectively, each sex plays to its own strengths. Indeed, the difference between the sexes is more pronounced in lions than in any other terrestrial meat-eating mammals. The bulk that serves the males so well in fighting to take a pride is a handicap when it comes to the stealth and speed required in hunting, where female lions are both swifter and more agile. The characteristic mane of the male, moreover, is conspicuous against the backdrop of the grasslands. Yet it would be wrong to conclude, as some have done, that males simply rely on the lionesses to provide for them. This belief was based on the greater visibility of hunts by lionesses, which often take place in open grasslands. More recently, however, the more secretive hunting strategies used by males have come to light.

In the early years of their bachelorhood, males do not have a pride to provide for them, so they have two choices—steal the kills of smaller carnivores, or hunt for themselves. In fact, they do both. Once they're within a pride, the males can and do exploit the kills resulting from the phenomenal cooperative hunting of the females, but they are themselves far less effective at this kind of hunting, to the point where joining in might handicap the females. Males come into their own in a different kind of habitat. By necessity they shun the faster prey that the lionesses pursue and specialize instead in animals that inhabit woodlands, particularly larger, heavier—and more dangerous—animals, such as buffalo. In this habitat, the males can conceal themselves more effectively and use explosive power rather than speed to make their kills.

Despite the difference in the approaches taken by the sexes, male lions are roughly as successful as females at hunting. Buffalo are formidable prey, standing nearly 6 feet tall at the shoulder and weighing up to a ton. They are not easily intimidated. They bring their own cooperative strategy to bear, turning to face approaching lions with a defensive barrier of lethal horns. To win out, the lions must panic the herd into fracturing, which involves brinkmanship on both sides as the lions feint at the buffalo line, trying to turn them. Not too close, though—a buffalo can easily gore a lion to death and is powerful enough to launch its persecutor bodily into the air. It's a deadly game and one that male lions are adept at. Their prize, if they win, is a gigantic feast, fit for the king of beasts. By following different hunting strategies and specializing in different prey animals, the males and females of the pride avoid competition with each other in a way that serves all well.

MANIC MEAT EATERS?

Among the large mammals with whom we share the planet, none is more reviled than the hyena. In East African folklore, witches (presumably small ones) are believed to ride on their backs, while West African mythology associates them with immorality and deviance. It's an image that has only been compounded by writers and filmmakers. In his book *Green Hills of Africa*, Ernest Hemingway described hyenas as "hermaphroditic self-eating devourers of the dead." In *The Lion King*, hyenas are cowardly, malicious, and untrustworthy. Even in nature documentaries, they're frequently depicted as thieves, parasites stealing the hard-won gains of nobler beasts. So hyenas sit alongside rats and cockroaches in the pantheon of undesirables. Yet in contrast to the other two, hyenas don't encroach on our domestic arrangements. In fact, given a choice, they would much prefer to avoid us altogether.

Why does the hyena have such bad PR? Partly it could be because of their dark, demonic eyes and ungainly appearance, not to mention their manic chattering, which has been likened to the laughter of a madman. And partly it could be the result of our need for a narrative. We view lions as representative of traits to which we aspire, so hyenas, the lions' greatest foe (aside from other lions), are cast as villains.

I have to confess that I find hyenas captivating. They were at the top of my wish list of animals I wanted to see in Africa. They live in

a unique society and exhibit some amazing behavior. Nevertheless, I
don't wish to attempt a whitewash. Like lions, hyenas have attacked
and killed people. In the wars that ravaged parts of Africa in the second half of the twentieth century, they readily consumed the corpses
of fallen soldiers. In certain tribal traditions, such as among the Maasai, the bodies of the dead were left for hyenas to take, and covered
in oxblood to encourage them, since having the body of a loved one
declined by hyenas was seen as a social disgrace. Perhaps in response
to an increasing familiarity with human flesh, and also as a result of
reductions in available prey, hyena attacks on living people have increased. The practice of sleeping outdoors during the hottest months
of the year places people at risk of opportunist attacks, as do nocturnal activities such as poaching. On rare occasions, a hyena may
attack people during the day. A local herder was mauled in Nanyuki,
near where I was staying at the time. The animal seized his arm, and
the man flailed at it with his free hand; since this produced no result,
the man's quick-thinking response was to bite the hyena's ear. The
biter bitten, it released the man's arm and retreated, no doubt with a
newfound, grudging respect.

Hyenas are not one species, but four. The spotted hyena, also
called the laughing hyena, is the largest, weighing in at up to 175
pounds. Alongside the spotted hyena are three other species—the
aardwolf, the striped hyena, and the brown hyena. The spotted hyena
is the most social and, to me at least, most interesting, so this is the
one I'll focus on. Spotted hyenas can be found throughout much of
sub-Saharan Africa. A glance at them suggests that they are related to
dogs, yet, surprisingly, they're closer to cats, and still closer to animals
such as meerkats and mongooses. The sloping back that is so characteristic of hyenas and that gives them the skulking appearance that people
apparently equate with shiftiness seems to be an evolutionary trade-
off. The powerful forelegs, shoulders, and neck provide the hyena
with strength for the attack and for carrying large hunks of meat—a
hyena specialty.

Their other specialty is their powerful bite—not for nothing are
spotted hyenas known as bone-crushers. They can crunch bones three

times thicker than your own thigh bone, which means that little gets wasted from a kill. Carnivores never know when the next meal is coming, so they tend to pack as much food in as possible when they get the opportunity. It's been suggested that in a single sitting, a hyena might eat as much as 30 pounds of food—roughly what a well-fed human would eat in a week—featuring a deliciously tantalizing mix of flesh and skin, liberally supplemented with bones and hooves, even horns and teeth if the animal is hungry enough. They've even been known to eat that crunchiest of snacks, the tortoise.

A Serengeti spotted hyena den in the daytime is a calm place. The ground is bare and dusty, with several tunnels leading down to a chamber where the young hyenas rest. That's not what they're doing today as I watch, though. Around ten adults are stretched out or sitting in the sun, a handful of cubs playing and wrestling with one another and occasionally teasing the adults. All in all, it gives the impression of a peaceful, relaxed society. It isn't always this way. As dusk begins to fall, the activity around the den increases. The adults are up and chattering in their strange, disconcerting manner—the clan is getting ready to move. These clans, like the lion prides described earlier, are the social unit of spotted hyena society. These are big groups, usually around thirty animals, though there can be as many as ninety, making hyena clans the largest social groups of any terrestrial carnivore, and indeed one of the largest of any mammal.

It isn't the size of the clan that makes spotted hyena society so interesting, though, but the unique and intricate relationships that occur within it. Female hyenas are usually slightly larger than males. (Among mammals, it's unusual for females to be the larger sex, but across the whole animal kingdom it's the norm rather than the exception.) In hyena society, the females are the most socially aggressive sex and entirely dominant—the lowest-ranking female almost always outranks the highest-ranking male.

Not only are female hyenas dominant, but they are also what is sometimes described as "masculinized." Specifically, females have what looks very much like a penis. It isn't actually a penis, but a "pseudo-penis"—a large, erectable clitoris. The labia, meanwhile, are fused into—you guessed it—a pseudo-scrotum. So both sexes have similar-looking apparatus between their legs. It makes it difficult to tell them apart, which gave rise to the idea that hyenas are hermaphrodites, or indeed, "sexually deviant"—another slander on their reputation. There's a story of a zoo that tried to breed a pair of hyenas for years before realizing they had two males. While it's not well understood how this anatomy first developed, both the penis and the pseudo-penis are very important in a hyena's social life. Both sexes are capable of producing a rather unusual signal, known as a social erection. The display of a social erection seems to have little, if anything, to do with sex. Instead, it signals submission.

While females are larger than males, rank within the clan isn't decided by size, or even by fighting prowess; instead, the clan is a network of alliances built on relationships and relatedness. It's rather like a feudal kingdom in some ways—the offspring of the aristocracy, in this case, a female cub born to the dominant female, automatically assumes a position one below her mother and above all the other females. Socializing starts early in the clan world. Cubs accompany their mothers and pay close attention to how they interact with other clan members. When a pair of hyenas greet one another, they do so in a carefully orchestrated manner in which the lower-ranking animal makes clear its subservience. Although hyena clans can occasionally erupt into violence, their social etiquette is fine-tuned to minimize aggression. After all, it would make no sense for the clan to spend all their time engaged in infighting.

Hyenas communicate their status to one another in all sorts of ways. Their calls, including the famous laugh, pass on information about an individual's identity and status, and their body language backs this up. Lots of things about hyenas seem back to front to us— that laugh, for example, is actually a signal of fear and nervousness. Subservient animals express fear when meeting dominants, curling their tails under their bodies and lowering and bobbing their heads. In some cases, the subordinate animal literally approaches the dominant one on its knees. Still, this is nothing compared to the so-called greeting ceremony that occurs when clan members are reunited after a period of separation. The pair of hyenas line up nose to tail, rather like dogs do when they meet one another, and sniff bums. The subordinate one of the pair lifts its leg first, allowing the dominant one to sniff its genitals and to see the conspicuous social erection that it produces for the occasion. Vulnerable as this makes them, hyena decorum demands it. If the subordinate animal fails to stick to the script, it can trigger vicious reprisals.

Watching her mother go through this process, a young female cub figures out where she fits into the social scene. If she's the offspring of the dominant female, it doesn't matter that she's small, because she has the support of the boss. Males will defer to her, and clan females, too, at least so long as the boss is around. In Serengeti clans, her mother may have to leave for hours or even days to find migrating herds of prey animals, and when this happens the younger animals are left at the den. Without the intimidating presence of the mother, other hyenas in the clan aren't necessarily so accommodating to her cubs and may take the chance to settle a few scores.

The way in which the dominant female assures her daughters a place at the top table in the hyena clan means that her group of kin females are very much like a ruling dynasty. Below them is the next most dominant female and her own daughters, and so on through each successive group of females until, at the bottom, are the poor, put-upon males. There are major benefits for the dominant female in this structure. She and her daughters get the best access to feeding opportunities at a kill, the most desirable resting places, and the best spots in the den. More food translates into more offspring as well as more milk for suckling young. Those young are healthier and grow faster than the young from less dominant animals, eventually having their own offspring at a younger age. They even live longer, so powerful is the effect of this rank precedence. By this process, the dynasty grows and strengthens. Meanwhile, sons born to the dominant female also benefit through their pedigree, although not to the same extent as their sisters. The sons of the dynasty are far down the clan pecking order, though they do receive some protection from their mother when they're contesting food, and their sisters are less aggressive toward them than to other males. And even though male hyenas don't play any important role in childcare, the regal daughters also tend to go easy on their dads.

As with most mammals, it is the young males who tend to leave home once they approach adulthood. Usually, they join a neighboring clan, where they have to take a place right at the bottom of the social pyramid. They're forced onto this path by the desire to breed. At home, in their birth clan, their chances are very limited; if they want cubs one day, they have to move on. A new male in a clan can work his way up, but he'll have to be patient, gaining status only as males above him die or leave. High-born sons seem to do better in their new clans than low-born ones, perhaps because of their well-fed and cosseted beginnings in life. Even so, it's tough being a male in the hyena world.

Hyena clans are a web of intrigue and alliances that help to support each animal in the struggle for status and power. To successfully navigate this complex social world, hyenas have a well-developed ability to recognize individuals. Beyond this, they have a sophisticated sense of the interrelationships between others in the clan. The importance of pedigree and relatives in high places means clan members are especially adept at recognizing their own kin—not only siblings and parents, unsurprisingly, but also more distant relatives, such as cousins.

How do they do this? Like us, and indeed like almost all vertebrates, hyenas possess a set of genes collectively called the *major histocompatibility complex*. This set of genes is important for determining disease resistance, but it has a side effect: it affects body odor. Every individual has a unique chemical signature. Some of that chemical signature is determined by diet and lifestyle, and some by genes. Because relatives share a proportion of their genes, they also tend to smell more alike than two individuals who are unrelated. Although our noses aren't generally good enough to distinguish these subtle differences, hyenas have an excellent sense of smell and can use the chemical signature of others to decide who is kin, and therefore worth helping, and who is not. Once they've identified their relatives, it forms part of a detailed dossier of information that they hold on other hyenas—not just who's who, but where they fit into the overall pattern of relationships among the clan.

All of this helps when the hyenas jockey for position and privilege. They build coalitions with other clan members that allow them to hold on to or improve their social status. The hyenas with the most allies are the most powerful. When they're deciding who to form alliances with, hyenas very sensibly try to ingratiate themselves with more dominant members of the clan. Having initially acquired their rank from their mother, young hyenas maintain a close relationship with her, and with the rest of their kin group, in order to reinforce their position. This coalition support goes both ways—the mother backs her daughters, and they, in turn, rally round her when trouble is brewing. Greeting ceremonies are an important part of this, reinforcing the bonds within coalitions as well as putting usurpers in their place.

As already mentioned, when males join a new clan, they enter on the bottom rung of the social ladder and initially have no allies. For a new boy, climbing this ladder relies on building relationships. Integrating is essential if he wants a chance at one day having his own cubs, but it takes time. In contrast to so many other animals, female hyenas have total control over their choice of breeding partners. They prefer males they've known for a long time—often years—and with whom they've developed a strong relationship. For a male, this means devotion: he may follow the object of his desire around for weeks before she consents to mate with him. There's no exclusive pair bond, though—the male has to keep working on his relationship if he wants to stay with the same girl.

Although the clan is the main unit of hyena society, clan members don't tend to stick together as a single group as they move through their territory. Instead, hyenas travel around on their own, or in small groups, before reuniting with others. There's a constant process of splitting and joining, making for a structure known formally as a *fission-fusion society*. In this lifestyle, hyenas often have to switch between independent wanderings and integrated pack hunting, which requires a good deal of social sophistication. The patterns of affiliation between hyenas are perhaps most apparent when there is competition within the clan, and particularly when they vie for access to a kill. Each coalition is like a gang within the clan world, and when times get hard, these relationships are especially important. Hyenas favor their relatives, who are typically fellow members of their coalition. They cooperate to get prime feeding access, even to the point of keeping other clan members off the kill, so it's clearly a case of who you know rather than what you know. Although younger hyenas generally get slim pickings, having to wait until the adults have fed, the cubs of the dominant female can sometimes piggyback off her status to get a bigger slice of the meaty cake.

The huge advantages of status and affiliation in the clan explain why hyenas invest so much effort into their greeting ceremonies and into slapping down potential challengers. They're also keenly aware of how relationships between other clan members shift and change over time. It seems as though hyenas, like us, aren't necessarily keen on change. When they see two clan hyenas fighting, they're most likely to join the fight on the side of the dominant animal, even if that animal looks like it's losing. They're cautious about intervening, though, as it isn't risk-free and might involve some loss of social prestige. In making this decision, they weigh up how close their ties are to the animal they might help. Close relatives receive help. How about more distantly related animals? Maybe not.

What is it that makes hyenas so aggressive? The answer seems to be hormones—more specifically, androgens, which regulate the development of male characteristics. Testosterone is the best-known androgen. Hyenas are born having been bathed in androgens throughout their development, and they're born crabby. Possessing a set of needle-sharp teeth and open eyes from day one, hyena cubs sometimes attack their siblings shortly after birth (scans of pregnant females suggest that cubs might even fight in the womb, before they're born). The result can be the death of one of the usual pair of newborns, especially when their birth coincides with a time of shortage. It seems strange for an animal who, in later life, is so reliant on kin coalitions, yet a single murderous cub is likely to get a double supply of milk.

Early exposure to massive levels of aggression-inducing androgens is probably what sets hyenas up for their belligerence in later life. What's more, the offspring of dominant mothers produce more androgens during their development than those of lower-ranking females, which might be a factor in helping these high-born cubs maintain their status. As they develop, the hormone patterns of males and females change. Just as in other mammal species, adult males have far more testosterone than adult females. This presents a puzzle. Typically, testosterone is linked to aggression, yet among hyenas, males are inevitably subordinate to females. If it's not testosterone that makes females the more aggressive sex, then what is it? The best guess is that it's caused by a few factors, including high levels of another androgen, androstenedione, and a quirk of hyena biochemistry that amplifies its effect. Add into the mix the parts of a female hyena's brain that regulate aggression, which are enlarged, and the result is female supremacy. Nonetheless, it's important to add that despite hyenas' supercharged aggressive tendencies, clan rivalries and infighting rarely result in serious injury. For such a powerful animal, hyenas show surprising control when meting out punishment, even side by side at a kill.

When the chips are down and the clan is threatened, these rivalries take a back seat. In Tanzania's Ngorongoro Crater, hyena clans have very clearly defined territories, and they don't tolerate intrusions by other hyenas. There's good reason for this—the territory contains the resources, especially food, that they need to survive. Loss of territory could lead to hardship or even starvation, so defense is a priority. When hyenas roam throughout their home range under normal circumstances, they tend to be either on their own or in small groups; but under threat of invasion, you can see groups of ten, twenty, or even more presenting a united front. Defense is a numbers game—the more the defenders, the greater the likelihood of winning. So different coalitions from within the clan come together, cooperating to defend their patch.

The boundaries between clans are understood by both sides and are clearly marked. Hyenas paste a fabulous waxy secretion, sometimes called "hyena butter," onto vegetation such as long grass stems. It's not the kind of butter you'd want to smear on your crumpets, though, not least because the hyenas produce it from their anal glands. This secretion is a kind of hyena calling card, containing information about an individual, such as its sex, dominance, and reproductive state. Crucially, it also contains a clan-specific smell, an olfactory proof of membership. Other hyenas pay plenty of attention to these scents and rub the paste of clan members onto their own glands. The paste contains a cocktail of delightful bacteria that contribute to its specific stink. By rubbing this paste onto their glands, the hyenas inoculate themselves, so that, over time, particular strains of bacteria pass to all clan members, giving them a common clan smell. Pasting occurs all over their range, particularly near the den. When they paste at the territorial boundary, they're acting like human gangs, tagging their turf with graffiti.

The effect of these territorial declarations at the boundary between two clans can be dramatic. Hans Kruuk, one of the great twentieth-century field biologists of the Serengeti and the Ngorongoro Crater, described a pack of hyenas in close pursuit of a wildebeest stopping dead when they reached the boundary, preferring to let the prey go rather than risk warfare by trespassing on the next-door clan. Males dispersing from their birth clan have to run this gauntlet if they want to switch groups. Resident hyenas treat single intruders with suspicion and hostility, yet such is the size of some of these territories that a lone male can sometimes travel far into enemy territory before he is detected. Once he has been spotted, the residents will approach him slowly, but an air of menace is obvious in their attitude. How the loner emerges from this depends on his body language. Nearly always, he will be a picture of submissiveness, tail and head down. Even so, he is likely to get chased and bitten. He'll have to be persistent if he wants to gain acceptance.

Small bands of hyenas patrol the borders of their territory. If they encounter a border patrol from a different clan, they may go through a series of greetings that allows the mini-armies to assess one another. The situation intensifies when a border patrol meets a hunting pack from their neighboring clan. The violence can be extreme, particularly when a kill has been made near the boundary. Both the females and the subordinate males join to attack intruders from outside their clans. Often, these skirmishes can pit males who have moved into a clan against relatives from their birth clans. Blood ties don't seem to restrict the males' aggressiveness, nor that of their offspring, who might be fighting against aunts or cousins. Once you're allowed into a clan, your loyalties shift. Clan over kin, in other words.

When times are hard, desperation can force clans to overstep territorial boundaries. The hyenas are well aware of the risks involved in turf wars. If they take down a prey animal in enemy territory, they will usually surrender the prize if they're caught by the residents, even if they outnumber them. Just being on the wrong side of the border seems to make them nervous. If they're not discovered, the trespassers eat rapidly and dismember the carcass so they can transfer it back over the border. But all of this takes time, and if the home clan finds them, things can get gory. Hyenas can be maimed and killed in inter-clan fights. Disturbingly, to us at least, these casualties don't go to waste—hyenas will cannibalize their own kind given the opportunity.

Hyenas are often portrayed as scavengers. It fits, or perhaps has played a part in forming, their long-standing image as morally corrupt and devious creatures. Hyenas certainly don't turn up their noses at carrion, and they will happily relieve other animals, such as jackals, wild dogs, and cheetahs, of their hard-won kills. They will even try their luck with lions. A familiar scene in nature documentaries is of lions feeding at a carcass, ringed by a pack of whooping and jabbering hyenas, pestering and harassing the big regal cats as they dine. If we could rewind the scene, as often as not it is the lions that evicted the hyenas from their kill. There is such a thing as a free lunch on the African savannah, and lions know it, homing in on the hunting hyenas in the hope of profiting from their efforts. A person arriving at a fresh kill where both lions and hyenas are present can, with practice, tell which side made the kill, because the two species dispatch their victims in different ways. Lions kill with a bite to the throat, and there are often telltale claw marks in the shoulders and flanks of the prey, made during the final stages of the pursuit. Hyenas, by contrast, bite at the hindquarters of the animal as it flees. And if the hyenas have fresh blood on their faces, then it's likely that they made the kill.

The much greater bulk of the lions means that, one on one, a hyena is hopelessly outgunned. The golden ratio for hyenas seems to be when they outnumber the lions by around four to one—then they have a chance of keeping, or seizing, a carcass. Most times, the lions keep a hyena pack at bay for hours while they eat their fill, snarling, lashing out, and sometimes chasing their hungry audience. The hyenas, very sensibly, keep a little distance between themselves and the lions, although they occasionally dart in to seize a morsel, if they think they can get away with it. Lions are known to kill hyenas, but seldom, if ever, do they eat them. Kruuk told of a hyena that was slow to recognize the approach of lions to a kill and, with its escape cut off, picked the only place it could think of to hide—it climbed inside the carcass and waited for the lions to sate themselves, and ventured from its grisly hiding place after they departed.

There is a large overlap between the diets of lions and hyenas, and consequently lots of competition between them. The established view of lions as hunters and hyenas as scavengers could hardly be more wrong: lions steal from hyenas something like twice as often as hyenas steal from lions. Hyenas are, first and foremost, hunters rather than scavengers. Yet a kill is a noisy event, and the maniacal laughing of the hyenas as they feed draws the attention of competitors for miles around. Why would hyenas make such a commotion when it seems against their interests? We can only guess; yet it's notable that, for such a voracious animal, hyenas rarely get seriously aggressive with each other over food, and it may be that their famous vocal display plays an important part in providing mutual reassurance. Hyena conversations at the carcass may also serve to attract other clan members, bolstering their numbers and making it more difficult for lions to seize the plunder.

In any case, hyenas eat very quickly. A group of hyenas may dismember and consume a large prey animal, such as a wildebeest or zebra, in less than half an hour, leaving no more than a patch of blood on the ground. Perhaps mindful of the risk of having their kill stolen, they carry away chunks of the carcass, either to eat them farther away from where the kill occurred or to cache them for later. Sometimes hyenas store their food for a short while in water, where it remains relatively cool and is hidden from the attentions of most other predators.

Hunting is a numbers game for hyenas, and they approach the task in a less obviously coordinated way than lions. Wildebeest is one of their favorite meals, and, when hunting them, their strategy seems to be to pick out the vulnerable ones in the herd. During the day, grazing wildebeests are scattered loosely across the plains. Hyenas move among them, inspecting and weighing up their options. Strangely, the wildebeests generally don't seem too perturbed about close encounters with this mortal enemy, seemingly able to judge when the hyenas mean business. Often, the hyenas can move within a few yards of a wildebeest and receive only a baleful stare in response.

But when it comes to winnowing out a victim, the balance changes. A hyena rushes at a group of wildebeests, who initially trot away, stiff-legged and alert, though not at full pace—yet. The hyena stops to assess the result, and it may perform several rushes before it picks a target. Hyenas appear to be keenly aware of differences between the wildebeests. Like champion poker players analyzing their opponents, they scrutinize their behavior for subtle signs of weakness. If a rush reveals a likely target, and if they can separate it from the herd, the hunt is on. Seeing this, other hyenas join the pursuit. An isolated wildebeest is now in deep trouble. Hyenas are fast, and they have tremendous stamina. Their hearts, proportional to their size, are nearly twice as large as a lion's, and they can run and run. The best chance for the wildebeest is to head for a dense group of its own kind. Failing this, it may be pursued relentlessly for up to a few miles.

The longer the chase, the better the hyenas' chances. The hyenas keep pace with their victim, snapping at its hind legs as it runs. Eventually, the wildebeest, brought to the edge of exhaustion, slows. Once the hyenas begin to overhaul it, what follows is nature at its very rawest. With their powerful jaws the hyenas tear at the animal's hind legs, rip away its vulnerable udders or testicles, and attack its soft belly and leg muscles. As they pull the wildebeest down, it disappears underneath them, succumbing to its fate—to be eaten alive, piece by agonizing piece. Kruuk described how the wildebeest seldom makes more than a token attempt at any kind of counterattack at this stage. The likelihood of success in any case would be remote; for the most part, the victim seems to be in shock as it falls.

When hunting wildebeests, it is usually a single hyena, or a small group, that starts the action, testing the prey by rushing at it before the activity evolves into a pack chase. But when they pursue zebras, their strategy is different. A group of ten or more hyenas set out with a more concentrated force. At up to 900 pounds, zebras represent a greater challenge for hyenas than the smaller wildebeests, which only weigh about 200 to 350 pounds. In addition, zebra, and in particular zebra stallions, fight back. The preparations for a zebra hunt are more coordinated than for other types of hunts, starting even before the hunters leave the den site. This time, they'll congregate at the den site and set off as a coherent group; when they hunt smaller prey, they tend to be less clearly organized. This difference in procedure suggests that the hyenas have some plan in mind for hunting a specific prey animal prior to departure. They can rove long distances when hunting zebras, going on journeys of dozens of miles, apparently with the single-minded purpose of finding a herd. The hunt itself shapes up in a similar way as with the wildebeests, however. The aim of the game is to isolate an individual and then run it down. Interestingly, some members of the clan seem to specialize in the actual hunting, while others follow behind. The killers get first pick of the food, while older clan members and cubs, who can't run so fast, join in after the kill has been made. On occasion, especially during the wildebeest calving season, or when among smaller prey, such as Thomson's gazelles, the killer hyenas will go on a spree—killing their victims, but not eating them straightaway. We see similar behaviors in foxes in chicken coops, or in domestic cats with mice. The instinct to kill is independent of the need for food. In hyenas, this surplus killing might benefit the clan as a whole by making large amounts of food available.

If all of this seems to paint hyenas as hyperaggressive, sibling-murdering, cannibalistic monsters, well, there's some truth in this. Nevertheless, to consider them only in this light would be to do them a grave injustice. While they are exceptionally effective predators, they have another side to them that is far less often presented. In the few instances where hyenas have been tamed, or at least habituated to humans, they interact with people with warmth and gentleness. As he conducted his fieldwork, Kruuk kept a tame hyena whom he named Solomon. Solomon traveled with him and slept beside him in a tent. Eventually, Kruuk donated him to the Edinburgh Zoo, because once he got a taste for cheese and bacon at the game lodges, he was insatiable. Solomon would batter his way through doors to get his fix from the buffet, much to the horror of dining tourists. Hyenas maintained at research institutes studying their behavior develop strong bonds with the researchers, greeting them with affection. Just like us, hyenas are highly social animals, and they are exceptionally good at building relationships. Like many other social animals, they are intelligent—remarkably so, in fact. Their sophisticated society, with its complex interrelationships, suggests that in terms of brain-power, they're up there with primates.

In some contexts, hyenas even outperform chimpanzees. When they were given a task to solve that required cooperation in order to get a reward, hyenas aced it. The task required two hyenas to pull a rope. To win the prize, they had to coordinate their actions, so that they pulled their ropes at exactly the same time. Although many monkeys and apes are highly social, most do not forage with the same level of cooperation as hyenas, which makes this kind of task difficult for the primates. The hyenas worked the problem out quickly without having to be trained for it. This result confounds the image of hyenas as dull-witted and devious; indeed, nothing could be further from the truth. As a final refutation of the hyena's entrenched and negative reputation, I offer Hans Kruuk's account of watching a group of hyenas at play—jumping into a river, splashing and fooling around, and generally having a riotous time. We do hyenas a great disservice if we only accept the clichés about their character.

RUNNING WITH THE PACK

It's early spring in the Canadian north. A trickle of water can be heard beneath the stream's crust of ice—the annual melt is beginning. Yet it will be several weeks before the land fully casts off the grip of winter. As dawn breaks, a woodsman emerges from a rough cabin. He pulls his coat tight around himself and checks his gun, then starts out to check the traps he laid late yesterday, his boots crunching through the frozen rime of snow and ice. He makes his lonely way through the stark country, keen eyes picking out familiar landmarks among the trees. The first trap yields nothing, nor the second. He moves on, but as he does so, a movement catches his eye. Coyotes? Wolves? He slides his rifle off his shoulder and pauses for a moment, holding his breath and peering into a stand of birch. All is still. Reassured by the rifle's weight in his hands, he sets off again. A few more paces, then another movement draws his attention. A wolf emerges from the trees, in front of him and to his left, perhaps 600 feet away. The animal holds no great fear for him; still, he slows his pace. Can he close the distance to allow a clear shot?

The wolf is not alone, and she and her pack are maddened by starvation. It has been a desperate winter. Another wolf, her son, steps out of the cover of the woods beside her. They are cautious, for their experience with humans demands it; yet their caution is tempered by their empty, aching bellies. The she-wolf looks back among the trees. The rest of the pack is gathering. She begins to move, trotting along the margin of the woods, flanking the hunter. Suddenly, a bullet smacks into a tree just behind her, followed a fraction of a second later by the sound of the rifle's report.

Missed! The woodsman curses and reloads with practiced efficiency. When he looks up again, he's surprised to see that the wolf, no, wolves, have not fled as they usually do. Instead, more come into view. Four, six, twelve now? More. He begins to feel uneasy as he lifts his rifle once again to sight them.

The wolves pause at the sound of the rifle, though none have turned tail. The she-wolf takes a pace toward the man, then another. The pack, emboldened by her courage and by their keening hunger, follow suit. They gather pace, and the distance between themselves and the man closes.

The woodsman fires another shot. This one is sure and a wolf cartwheels over, dead before it hits the ground. The pack show no hesitation. On they rush. Another shot, and another, and another. All are true. Four wolves are down, but still they come. He shoots three more before they are on him. One grabs for his thigh, and he smashes at it with the stock of his rifle, barely registering the thud as its skull gives way. Now two more are tearing at his legs. Desperately, he brings the heavy stock down on each of them and then swipes at a third, catching it hard on the jaw. There are too many. A huge wolf launches at his chest and he is brought down; others seize his arms, his legs. Though he punches and kicks, he is hopelessly outnumbered. He feels the power of his foes and the tug of their jaws, suffering no pain, only shock. For the wolves, a man is a meager meal, and a costly one. When other men come along some days later, they find a grisly scene. The carcasses of eleven wolves, seven shot, four clubbed to death, surround the woodsman's remains.

This terrible fate befell a woodsman, originally named as one Ben Cochrane, near Lake Winnipeg in Manitoba almost a hundred years ago. A few weeks later, the real Cochrane turned up, though: hale, hearty, and demonstrably unchewed, and doubtless surprised to discover the furor. The true identity of the victim remains unclear. For all the awful detail of this incident, however, wolf attacks are rare. When they do happen, it tends to be because the wolves' instinct to avoid people is overcome, either by extreme hunger or because of a kind of madness induced by rabies. In other cases, when wolves live for extended periods of time close to human habitation, they may become habituated to people and lose their fear of us. Attacks are now far less frequent than they once were, partly because wolves have been driven out of much of their natural range, and partly because they have learned to avoid humans and the guns they carry. Nonetheless, the representation of the wolf as the "four-footed fiend of winter" lingers in our ancestral memory, especially in the folklore of cultures in the north. That seems strange when you consider that we can be so affectionate with the domestic dog, their close relative, which has earned a reputation for faithfulness and companionship. The relationship between dogs and their owners has its roots in the very same sociability that characterizes the lives of wolves: these are, like us, grouping animals that form close bonds within their pack. The bonds that led to pack hunting are the selfsame bonds that led to the rapport we have with our more familiar canine friends.

When we talk about wolves, we often do so in a way that implies one single, identifiable species. Yet even with the molecular tools that biologists use nowadays to identify and distinguish species, debate still rages as to which ones are distinct species, which are subspecies, and how they are all related to one another. All wolves belong to the family Canidae, also known as the dog family, and to the genus *Canis*, which includes jackals, dingoes, and the domestic dog as well as wolves. There are three different species of wolves: the red wolf, the Ethiopian wolf, and the largest and most widespread variety, the gray wolf. So far, so good. But the gray wolf is split into lots of different subspecies (or even into separate species, according to some experts).

These can interbreed both with each other and with domestic dogs, which further blurs the distinction between all of them. Not only that, but you could be forgiven for thinking that a gray wolf is, well, gray. It's not, at least not always—gray wolves can be gray, but they can also be white, black, brown, or red. Gray wolves are also known by a variety of different names—common wolf, timber wolf, plains wolf, tundra wolf, and so on. You start to get an idea of why there is so much controversy.

The gray wolf, in all its forms, once had a vast distribution. Its range stretched across the Northern Hemisphere from North America to Greenland and from Europe through Asia, as well as south to India and east into Japan. Over the centuries, humankind pushed it out of many areas. Remnants cling on, particularly in cold, remote parts of the globe. It is in these areas that this emblematic and—to some—terrifying animal may still be seen close at hand. It isn't easy to find them, though, particularly in mainland Europe. Just south of Berlin, in what was once East Germany, there are large swaths of countryside where Soviet forces used to undertake military exercises. Civilians were excluded from the area, and, even now, rumors of discarded live ordnance mean that much of this land is deserted. Although wolves were thought to have been eradicated from Germany at the beginning of the twentieth century, packs have crossed from nearby Poland to establish territories where the Red Army once drilled.

I traveled to the area in the hope of seeing the wolves, yet despite records of a thriving wolf population, my efforts were unrewarded. I found signs of wolf activity, including wolf footprints and spoor, but the wolves remained out of sight. Still, I had a sense that they saw me, and that they were nearby. In Poland, a friend of mine did come across a wolf. He was researching another European mammal recovering from centuries of hunting, the bison, and spent months in the primeval Białowieża Forest. In a scene reminiscent of the lives of past generations, he was carefully moving through dense woodland when he emerged into a clearing and came face to face with a wolf. For a few moments, each watched the other in perfect stillness, before the wolf turned away and slowly retreated to the undergrowth.

Although the image of wolves as remorseless and dangerous killers is fixed in our psyche, chance encounters like this, ending in a retreat, are far more common than attacks—at least they are nowadays. Huge numbers of wolf attacks have been documented historically. In France, records show over five thousand attacks from the fifteenth century to the early twentieth, when wolves were virtually wiped out from the region. Yet, in the whole of the wolf's stronghold from North America to northern Europe and Russia, there were only eleven fatalities in the second half of the twentieth century. There are far fewer wolves now, of course, but undoubtedly, changes in human population patterns—with more people living in towns and cities—combined with a better understanding of wolves to help reduce the number of attacks. Wolves are one of the most well researched of all animals, and this has inevitably given rise to an increased appreciation for them, not least for their social behavior.

A wolf pack is often a family, parents and pups working together, although two or more families sometimes band together into larger packs. There's a strict hierarchy. A mated pair of wolves, the *alphas*, boss the rest around and are the only ones allowed to breed. While they aren't always the largest wolves in the pack, the alphas are usually the toughest; they have to be to withstand challenges to their authority. Dominant animals signal their social superiority with their body language. A wolf with its head held high, ears pricked and tail held out, is communicating that it is a high-ranking individual. Other pack members defer to an alpha by adopting submissive postures, such as by approaching at a crouch with tails curled beneath their legs, before licking the dominant wolf's nose and snout in a gesture of supplication. Throughout this procedure, the alpha maintains a cool dignity, accepting the fawning while staring straight ahead. At other times, lower-ranking animals may roll over onto their backs in a show of subservience and vulnerability. If a subordinate wolf fails to pay homage with these displays, fights can break out. Rivals will snarl and bare their teeth in an unmistakable threat display. A fight is to be avoided, if possible, so they attempt to settle the issue with a series of aggressive ritual displays, such as snapping their teeth together and lunging at one another without making contact. Only if these measures fail do fights break out, but when they do, they can be extraordinarily vicious. If a challenger overcomes an alpha, the rest of the pack members may turn on their former leader, mobbing and attacking it and driving it off. The deposed alpha may be lucky to escape with its life.

Although every wolf is aware of its status within the pack, transgressions do occur. Lower-ranked females may try to seduce the alpha male, or subordinate males may try to breed with the alpha female. To prevent this, the alpha of each sex must keep a close eye on the behavior of the other members of their own sex. But the urge to breed is powerful, and despite the danger, sneaky alliances happen. Within the larger packs, a certain proportion of the young are not the offspring of both the ruling alphas. In most cases, they are the offspring of the so-called *beta* wolves—those next in line in the hierarchy to the alphas, and the ones most likely to challenge them for supremacy. The wolf right at the bottom of the pack's social order is known as the *omega*. These individuals are subject to terrible bullying and, if food is short—and the alphas vindictive enough—may even be excluded from access to a kill. Oddly, the omega seems to play an important role in cementing the pack together, since having someone on whom to vent aggression allows a more peaceful coexistence for all. Another part of the social glue that holds packs together is wolves' playfulness. Mock fighting and chasing are all part of day-to-day life in the pack. They invite one another to romp around with a kind of bow, their hindquarters raised, tail raised and wagging, and front legs and head low to the ground. During these times, status seems less important than it usually is and even omegas can end up chasing their pack leaders around as all submit to the fun of the game.

Wolf pups gain the support of the pack and of their parents, the alphas, during the early stages of their lives. Once the pups are weaned and eating meat, alphas sometimes forgo their right to forage first at a kill, to allow the youngsters to feed. Competition for food is nevertheless often intense, and having another adult to feed will stretch resources to the limit. Youngsters will eventually have to seek their own mates, and to do so they may have to look beyond the close relations that make up their pack. When they're on the cusp of their adult life, wolves may thus leave their home—sometimes they're driven away—and their options at this time are limited. It is rare for loners to join an established pack—indeed, they run the risk of being killed if they stray into defended land. Without the protection of their kin, they are vulnerable, and many struggle. Even if they survive, their wanderings may last for weeks or months, during which time they may wander hundreds of miles from their birth pack. They're seeking a mate, often another lone wolf, and a patch of land that they can claim for their own. In areas with large wolf populations, the young loners must pick their way through a mosaic of pack lands, treading cautiously to avoid confrontation with the residents. Then again, amid this danger can lie opportunity. As the lone wolf skirts the territories of other packs, it may be able to entice one of that pack to mate. Alternatively, the loner may meet another wanderer as it travels, and this wanderer may be a suitable mate. Once they meet up, they can establish a pack of their own.

The key to survival for a wolf pack is defending their territorial claim along with its hunting rights to the prey animals that live there. Like lions and hyenas, owners of a territory will advertise their presence with scent-marking, especially along pathways at the margins. They reinforce these signposts in the concert of enthusiastic howling for which they are so famous. The nightly howling lets neighboring packs know that the owners are in residence, as well as conveying information about the size of the territory. (It's for this reason that lone wolves seldom howl.) Nevertheless, a claim can be disputed, and territories frequently come under threat. Maintaining ownership, or even capturing another pack's territory, is critical if the pack is to survive and thrive. Strength comes with numbers—border skirmishes and outright war between neighbors are often decided by the greater fighting ability of the larger group.

Disputes between packs are fiercely contested, and casualties can be high. In many areas, the greatest killers of wolves are other wolves. Yet even with the aggression that occurs within or between wolf packs, wolves also show strong signs of the kinds of affiliations—we might call them friendships—that we can recognize in our own social groups. Seasoned wolf watchers have described a sense of mourning within packs when one of their members dies, to the point that the pack may not hunt for a short time. As a scientist, I have to be cautious when trying to understand the emotional state of an animal, but we can get some insight into an animal's stress levels by measuring how much cortisol is in its blood. These measurements in wolves following the loss of a pack member do show a stress response, so perhaps these tales have some truth to them.

FOUR-LEGGED FRIENDS

Outside Shibuya train station in Tokyo, there stands a statue to a dog. The monument commemorates one of the most enduring and endearing stories in the long history of the relationship between humans and our canine friends. Each day when Hidesaburō Ueno returned home from his work at the University of Tokyo, his dog, Hachikō, would travel to meet him at Shibuya Station. Tragically, only a year after adopting Hachikō, Ueno died. Yet for the next ten years until his own death, Hachikō would return faithfully each evening to the spot outside the station where he used to meet his master. Although the extent of Hachikō's devotion is extraordinary, dog lovers the world over have firsthand experience of the amazing selflessness and affection of many dogs. Among these dog owners, the suggestion that wolves mourn lost pack members comes as no surprise. From the dog's perspective, the owner is a member of its pack. Dogs, like wolves, are essentially pack animals.

How similar is the behavior of dogs to that of wild wolves? To answer this, we need to go back in time to uncover the origins of modern dogs. The partnership between dogs and people has a long history; indeed, it is the longest relationship of its kind that we have with any animal. Evidence from archaeological digs suggests that dogs have lived alongside us for at least fourteen thousand years, and possibly far longer—perhaps over twice as long. Excavations of ancient human burial sites have turned up the remains of dogs interred alongside their masters, hinting that these were companion animals. At the time that our own ancestors and our dogs' ancestors first forged a partnership, human civilization was in its infancy. Most humans were living as hunter-gatherers, as it was before we made the transition to growing and rearing our own food. At first sight, it seems an unlikely alliance—two highly developed predators that normally would present a major threat to one another, and exist in a state of mutual fear and aggression, instead becoming the best of friends. How did these two species overcome that enmity to become so deeply intertwined? It's been suggested that the best candidate for early domestication would be a predator, since they are naturally less afraid of other species, and that it would also be a social animal, and thus predisposed to communal living. And wolves fit the bill on both counts. But the ancient history of the partnership between our species is still shrouded in mystery. The best we can do is to construct plausible scenarios based on limited evidence.

One suggestion is that wolves living at the edges of human settlements learned to scavenge food discarded by people. Wolves, like so many other animals, have distinct individual personalities. Some are inclined to be highly aggressive, others far less so. The people who lived in those settlements would have been more likely to tolerate the wolves with a more docile temperament, while driving off their more ferocious counterparts. The docile wolves would therefore spend increasing amounts of time near the settlements, where they would lose their fear of humans and become even tamer. Benefiting from the reliable food supply, over successive generations they would have bred lots of tame pups. Clearly, the friendliest wolves in the population would have gained an advantage from teaming up with us, and our ancestors, in turn, would have gained a ready-made guard and a useful hunting companion. Over time, the ties between our two species were strengthened, and these wolves became increasingly distant from their fully wild cousins. Although we might think that we domesticated wolves, it's probably more accurate to say that they domesticated themselves.

While this version of the wolf domestication story has gained many followers, it has its critics. The detractors argue that the humans of those times would have been cautious about discarding food, precisely because it would have attracted unwelcome attention from animals such as wolves or bears. And even if they did throw away food, these leftovers would be unlikely to satisfy large animals like wolves. Finally, throughout history, human attitudes toward scavengers have seldom been particularly welcoming. Would our ancestors have invested in a relationship with garbage dump raiders?

An alternative to the scavenger hypothesis is that the bond between wolves and people was forged through more complex factors at work in our millennia of coexistence. Essentially, this explanation points out how humans and wolves lived on the same land, shared resources, and over time learned from each other. While familiarity may often breed contempt, it could also have fostered a mutual respect between our two species that eventually led to tolerance and cooperation. Some support for this idea comes from studying the attitudes of indigenous peoples. In North American and northern Eurasian hunting cultures, wolves are regarded with respect and even reverence. Encounters between wolves and humans during hunts or at carcasses give rise to a need for understanding. At first, this is essential simply to avoid injury from a dangerous competitor. But perhaps over time it morphed into collaboration between our two highly social species. Working together is a powerful strategy that could benefit all.

We may never know the truth of how wolves first came to be domesticated—it's an archaeological jigsaw puzzle with many missing pieces—but thanks to a remarkable program of experiments carried out in Russia in the second half of the twentieth century, we have an insight into the process of domestication. Starting in 1959, the scientist Dmitry Belyaev ran a selective breeding program in foxes. For the purposes of his experiment, he chose foxes on just a single behavioral trait—tameness, scoring the foxes on how they initially responded to humans. He took the foxes that were most willing to approach an experimenter, and that showed the least fear or aggression when near people, and bred them. Then, in each successive generation, he tested the foxes again, breeding only the tamest. Every good experiment needs a control to provide fair comparison, so alongside his selective breeding program, Belyaev also chose some foxes at random for breeding. The two populations were otherwise kept in identical conditions. In between tests for tameness, Belyaev was careful to avoid interacting too much with the foxes—he was keen to avoid training them to get used to people. After just three generations, Belyaev started to see solid results—his foxes were becoming ever more tame. The proportion of tame foxes increased in each generation, and after twenty generations he judged one-third of them to be tame. After thirty generations, it was up to one-half. The experiment went on for decades, and by the early years of the twenty-first century, all the foxes in the breeding program from the tame group were, in effect, domesticated. By comparison, the foxes in the control group were much the same as they had been at the outset of the program.

Of course, this in itself isn't so surprising—we know that breeding for a particular trait increases the proportion of animals exhibiting that trait. The intriguing thing was that Belyaev's tame foxes had changed in other ways, too. They didn't just tolerate people; they behaved like domestic dogs. They were more playful, wagged their tails, licked people's hands, and competed for attention from their human handlers. They even developed different colored coats, shorter muzzles, smaller teeth, and floppy ears. Incredibly, all these changes came about simply as a by-product of breeding individuals with a friendlier disposition toward humans. The unexpected traits such as floppy ears, so familiar to us from our experience with domestic dogs, seemed to be linked: select for tameness and the rest of these characteristics come along as a package. It's reasonable to imagine a similar process taking place with wolves all those millennia ago, a process that ultimately led to dogs as we know them today.

As Belyaev's foxes became friendlier, another revealing change occurred in them as well—they became ever more attuned to understanding human gestures. This wasn't something Belyaev had set out to achieve; nor did it develop through familiarity with humans. It was just part of the suite of changes that developed alongside tameness. In fact, the foxes became as good at reading people as domestic dog puppies. This result was especially impressive because dogs are exceptionally good at this—far better than wolves, and better even than our superintelligent ape cousin, the chimpanzee. As dog owners, we often take it for granted that we can point them in the direction of a ball that we've thrown, or that they can pick up on subtle changes in our moods or behavior. Dogs are so attuned to us that when their owner yawns, they yawn, too, particularly when they have a very close relationship with the owner. The skills that these animals have evolved to help them navigate interactions with others of their own kind have broadened to include us. We've become part of the dog's social world, as they have become part of ours.

CHAPTER 8

CODAS AND CULTURES

Whales and dolphins are among the most enigmatic—
and most cooperative—of all social animals.

AN ENCOUNTER

Legs trailing in the water, I grip the side of the boat, alert for my cue. Rising to the top of a swell, the skipper spots something in the distance and kills the engine. "Go, go, go!"

I drop from the side of the boat with my buddy into the Atlantic Ocean, its floor thousands of feet below us. A moment later, the boat is gone. We're alone. The water is as clear as the air above us; I feel a sense of vertigo as I float above the abyss. I quell this irrational thought, training my gaze on the watery horizon. All I can do now is wait, and hope. Then, a huge shape appears against the blue, on the edge of my vision, and another, and another; gradually the shapes become more distinct as they head straight at me. I hang in the water, electrified by the sight—three of the largest predators on the planet, the fearsome protagonist of perhaps the most famous seafaring novel of literary history, *Moby-Dick*. I am face to face with sperm whales.

I was in the Azores to study the social behavior of whales and had gone there with some trepidation. Although this group of mid-Atlantic islands has a resident population of sperm whales, making it one of the best places in the world for biologists to study these majestic animals, the Azorean relationship with whales has not always been a harmonious one. Whaling was for many years an important part of the culture here, continuing until 1984. Sperm whales, like many of their relatives, are long-lived animals, with a life span comparable to that of humans. Although twenty-seven years had passed since the end of whaling in the Azores, it was likely that adult sperm whales in the region would have experience of humans as hunters. Reason enough, I pondered, for these intelligent beasts to be cautious, or even aggressive, when encountering us in the water.

Nonetheless, the chance to encounter the largest toothed whale in its own realm was too good to miss. This was all the more the case since getting the permits to jump in with the whales is not easy. Who knew if there'd be another opportunity? Still, the image of Gregory Peck lashed to the harpooned leviathan at the climax of the 1956 film of Melville's famous novel played in my mind as our team of four headed out from the Port of Madalena for the first time. On that particular trip and the ones that followed over the next few days, we had only the most tantalizing glimpses of the whales as they disappeared into the blue. The small boat we were using was maneuverable, but it didn't cope well with the large swells, and finding whales is challenging in rough seas. It did give us a crash course in dealing with seasickness—I took an epic amount of drugs to stave off the *mal de mer*. One of my companions, Romain, decided his body was a temple, so eschewed the chemicals, with the result that he spent most of his time draped over the side of the boat fervently wishing he were dead. Each day of that initial period was a carbon copy of the last, washing up and down the Atlantic rollers, eyes peeled on the horizon, the only soundtrack Romain's periodic, heartfelt retching.

We racked up hour after hour of zilch, but that's all part and parcel of looking for animals—if you want guarantees, go to a zoo.

Our prospecting was aided by the gimlet eyes of an ancient mariner, Joao, employed as a lookout and ensconced in a hut halfway up the volcano that originally gave birth to Pico Island. Strange to think that Joao had learned his trade and sharpened his skills by being the spotter for whalers years before. Times had changed, even if his job hadn't. We came to understand the whales, rather than slaughter them. But for four days, even the experienced Joao struggled to spot any whales in the surging sea. The telltale sign of the whales is their spout, the steamy exhalation of air and bits of other, less agreeable things that are fired out of its blowhole at the end of a dive. A decent-sized whale might launch its gust of moist air above the surface, but amid rough seas, you still need good luck to detect it.

Far below the waves, the whales were feeding. They're prodigious divers, capable of descending more than a mile into the darkness of the midnight zone for more than an hour at a time. Generally, however, they don't need to push themselves so hard—it all depends on where they can find their food. Sperm whales are prodigious hunters and can easily munch their way through half a ton of squid and fish in a day. There's little, if any, light in the depths of their feeding grounds, so they're heavily reliant, like oversized bats or dolphins, on echolocation to detect their quarry. That said, many species of deep-sea squid are bioluminescent, producing pulses of light as they communicate with one another and hunt. Having prey that obligingly illuminate themselves amid the darkness can aid the whales, but by the same token, being surrounded by a constellation of flashing squid can be disorientating. It's little wonder that canny sperm whales have learned to approach fishing vessels and pluck hooked fish from long-lines, though this is more a bonus for the whales than an effective way to satisfy their gargantuan appetites.

To tip the balance in their favor, especially when hunting larger and more elusive prey, sperm whales coordinate and cooperate. They descend to their feeding grounds in pairs or small groups to form a search cordon, a line of whales spaced over half a mile of ocean, a smart solution for locating clusters of prey. Finding a dense patch of squid is only part of the battle, however. Traces taken from underwater GPS devices mounted on the whales show that they divide to conquer—one whale dives below the squid to cut off their escape to deeper water, allowing the other whales to attack the flanks of the prey group. Nonetheless, our understanding of their hunting, like so many aspects of sperm whale behavior, is in its infancy.

Prey don't get much more fearsome than the giant squid, thought to have been the inspiration for the legend of the monstrous kraken: more than 30 feet long, they are not much smaller than the whales themselves. The heads of older sperm whales are often adorned with huge circular scars, the imprints of the giant squid's suckers, testaments to a history of struggles between these marine behemoths. Autopsies of whales have recovered the remains of squid in their digestive systems, confirming that they do eat such daunting prey; what isn't known is how they subdue such monsters. The lower jaws of sperm whales are surprisingly delicate in appearance, and though they're studded with long, conical teeth, there are even examples of toothless elders managing to forage. Furthermore, large squid recovered from the bellies of whales sometimes have no toothmarks in them—they seem to have gone down without a fight. Based on these intriguing details, it was suggested at one point that the sperm whales applied a literally stunning tactic to overwhelm them. The outsize head of a sperm whale—specifically, its fatty melon, located in its forehead—acts like an acoustic lens, focusing the sounds that the whale produces and intensifying their volume. Armed with this natural weapon, perhaps whales could produce a sonic boom to stun their prey. It made perfect sense, until it didn't. Lab experiments testing whether the kinds of noises the cetaceans produced could incapacitate other marine animals failed to show an effect, and, more recently, recordings of hunting sperm whales picked up the buzzes and clicks of echolocation but no giant booms. For now, the riddle of this charismatic whale's supremacy over its most formidable adversary remains part of its mystery.

Such fulminations filled my thoughts as the storm kept its hold on the Azores. Finally, on the fifth day of our trip, the waves relented. At last, we had a chance. Sure enough, it wasn't long before we heard the radio crackle into life and an excited voice reeling off directions in Portuguese. The skipper shifted course and told us that a pod of sperm whales lay just over a mile to the northwest.

The previous few days had at least allowed us to plan the protocols of getting in with the whales. The skipper would chart the path that the whales were moving along and then drop us into the water several hundred feet ahead of the whales before taking the boat well out of the way. And then we would simply wait, or perhaps fin toward some landmark to give ourselves the best chance to intercept and observe the whales. If the whales decided to change course, or to dive, then that was just bad luck. If there were to be encounters, these would be entirely on the whales' terms. Moreover, a whale cruising calmly along travels at a pace far greater than anything a biologist can manage, even with fins. Our meetings with the whales in the first few days of the trip had been infrequent and brief. The whales passed us by at the limits of our vision, or far below us, rolling onto their sides to peer up at us with their surprisingly small eyes, then they were gone. So we expected, at best, a few precious seconds with the whales as they passed, enough, if we were really lucky, to notice a few things, such as identifying marks or scars.

But this time was different. It wasn't only the ocean swell that had calmed—the whales, too, seemed in less of a hurry. Rather than cruising past, they lingered, and suddenly we found ourselves at the center of a family frolic. It was a phenomenal experience, greater by far than I'd dared to dream of. I couldn't simply hang at the water's surface and enjoy it passively, though; the cavorting whales kept coming perilously close, forcing me to scoot out of their way each time a mighty tail threatened to knock me spinning. The pod was made up of four whales—a huge matriarch more than 30 feet long, a slightly smaller individual about three-quarters her size, and two calves. Wonderful as all this was on its own, there was a cherry on our cetacean cake—with the pod was an adult bottlenose dolphin.

The two species are tolerant of one another, but their different life-styles and prey preferences mean that they seldom associate. What might have decided the issue was that the dolphin had a pronounced curvature of the spine, twisting its body just behind the dorsal fin. It didn't look like an injury (there was no scarring), but rather something the dolphin had carried since birth. Nonetheless, it had survived, against the odds, to reach adulthood. It's possible that the condition hampered its ability to swim at the relentless pace at which bottlenoses typically travel. If so, it would be isolated from the intensely social life of its own kind, and perhaps, as a surrogate, it had joined the whale society.

For the next twenty minutes, the whales kept up a constant dialogue with one another, making their ethereal creaking, knocking, and clicking sounds, while periodically the higher-pitched call of the dolphin could be heard. The whales rolled around in the waves at the surface, the smaller members of the pod circling the huge matriarch. Then, even more astonishingly, the whales began some strange kind of game. The matriarch would open her oar-like lower jaw, and one of the smaller whales would swim into her mouth, its head protruding from one side and its tail sticking out the other. The matriarch would then seem to very gently nibble the smaller whale for a second or two. The nibbled whale would swim clear and circle around to join the back of the queue, and another would maneuver into place for a little of the same treatment. The bottlenose joined in the fun as well, swimming into the matriarch's open jaws for its turn and receiving a toothy squeeze. I remained mesmerized by the encounter long after I'd left the whales to their play; it was an incredible privilege to get a close-up perspective of the remarkable social behavior of this little-understood animal.

Back on land, I pondered what it meant for the whales to be held in the matriarch's mouth for a moment. Maybe there's some parallel with the grooming behavior of primates. While the immediate role of grooming might be to keep the fur glossy and bug-free, more important is what underlies it, the act of building and securing relationships. Lacking dexterous limbs, of course, the whales can't do this. Perhaps this was their creative way of physically expressing themselves. Sperm whales live in matrilineal social groups, and the core of the group is formed by related females, often comprising a grandmother, her daughter, and their offspring. Sons live in these groups only as juveniles. As they approach sexual maturity, the males break from their social group and adopt a more solitary existence, although it's not unusual for males to form into loose bachelor groups with one or more other males. The group we saw that day was a fairly typical example of sperm whale society, so it could be that what I'd witnessed was some maternal attention being paid to the family in the form of a strange cetacean embrace. That the dolphin joined in suggested that it understood no threat was involved, while the fact that the matriarch lavished some attention on the dolphin suggests it was an accepted, though perhaps temporary, member of the group.

Although sperm whales are thought to have reasonably good eyesight, their primary means of communication is by sound. The aquatic environment sometimes poses challenges to visual communication, but water transmits sound far more effectively than does air, a fact that whales of many species use to their advantage. Among all the different sounds that the whales produced, one was particularly distinct, and they used it only when they first approached us. It was a thumping, thudding noise, strong enough that it seemed to penetrate your entire body, but not painfully so. The best way I can describe it is that it sounded like the noise you'd get if you hit a tire hard with an iron bar. There didn't seem to be any aggression behind it; it just seemed like the whales were curious about us, and exploring us both visually and sonically with these booming thuds.

Though nothing matched the intensity of the whales' sonic inspections, by far the most frequent vocalizations they produced were the rapid pulses of deep sonorous clicks and creaks. Each individual whale contributed to the conversation with its own series of clicks, known as a *coda*. The sound structure built into the coda is unique to the individual whale, which allows them to recognize one another by sound. Thus they manage to maintain contact far beyond the range of sight. Each different social group has a repertoire of codas in their communication, but a noteworthy facet of sperm whale society is that the social groups themselves belong to a larger and more loosely organized social structure known as a clan. While each social group usually consists of no more than ten individual whales, the clan may comprise several hundred or even thousands of animals distributed across thousands of miles of ocean. Each clan has its own unique dialect, producing clan-specific codas. This seems tied to its geographical range, in the same way that regional accents develop within language groups in humans. When sperm whales encounter one another, they can determine who they're speaking to as well as their family group and clan.

Being able to communicate is clearly important for the cohesion of animal societies, and it may be particularly important for the whales to locate one another when they return from a dive to the abyss. The very youngest calves, unable to dive to the prodigious depths their mothers can reach, remain at the surface while she patrols hundreds of feet below, searching for deep-dwelling squid. Newborn calves, which may already measure more than 10 feet in length and weigh about a ton, bear creases along their flanks for their first couple of days, apparently the result of being folded in their mother's womb. As with humans, the youngsters sound different from the adults, communicating with noticeably higher-pitched calls.

The presence of a newborn is significant not just for the mother but for the group as a whole. The calf relies heavily on its mother, especially for protection and for suckling, but it may also turn to another obliging female relative for a top-up of food. Like all mammals, the calves are raised on a diet of milk. They face an extra challenge, though: How do you drink when you're already in the drink, so to speak? The answer is that whale milk is like cottage cheese in texture—it's more a case of the calves eating it than slurping it down. I say calves, but even those nearing adulthood seem to enjoy a supplement from their mother, sometimes into their teens. Female relatives sometimes remain at the surface with the youngest family member on babysitting duties while its mother forages. This is not always so, however: the young whale may be left to its own devices while the mother's away. Like the young of other species, baby sperm whales are curious and playful. I found this out firsthand when a youngster sought us out as we lingered at the surface after a failed attempt to see a dolphin pod. Drawn to us for whatever reason, the little whale swam around us trustingly, nudging and nuzzling as it went. This was an extraordinary encounter in its own right, though potentially hazardous. It was obviously imperative not to stress out the calf, though it seemed to be enjoying playing its own variation on the game of tag with us. Another consideration was how the mother might react. If we roused the protective instinct of a 15-ton whale, things could go badly wrong. Yet the calf wouldn't leave us. If we swam away, it would follow, and, young though it was, it could effortlessly outpace our pathetic swimming attempts. We had no choice but to accept its attentions, nervously looking into the depths for its mother's return. When she did resurface, a few minutes later, she seemed unconcerned, and remained with us for a little while longer before escorting her charge away.

On the final day of the trip we made one last sortie to the whales. Luck was on our side—it had been four days since we'd first met the sperm whale group with the dolphin, and here they were again, the dolphin still very much part of their scene. Weeks later, after we had left this maritime paradise, we heard that our guides had again seen this group, complete with dolphin. This was a longer-term arrangement than I'd imagined; the dolphin was interacting with the whale social group to a surprising degree. If nothing else, it gave us some idea of the extent of the social tendency of both species, the deep-seated drive to seek and remain in company. In many ways, this unusual partnership raised more questions than it answered. For example, how did the dolphin manage to forage, encumbered as it was by its scoliotic spine? Based on its appearance, it was certainly well fed—like a barrel, in fact. It couldn't be foraging alongside the whales, because the dolphin couldn't match its adopted family's prodigious dives. Was it catching its own food? Or were the whales providing for it in some way? Sometimes, sperm whales bring their squid prey to the surface with them. Perhaps the dolphin was able to help itself to morsels. This seems a stretch, but however it was nourishing itself, the dolphin appeared to be an accepted member of the group. It's a demonstration of the unusual structure of sperm whale society that this could happen. Among many similar mammal groups, to be accepted into the fold you have to be a blood relation. While kinship is important to sperm whales, it's not the sole determinant of their associations. Genetic examinations of their social bonds reveal that they form long-term relationships with both family members and outsiders. Although the dolphin might have been taking this to an extreme, it suggested a remarkable flexibility on the part of both species.

ATTACK AND DEFENSE

Another remarkable feature of these highly social mammals is their response to threat. Until recently, it was confidently stated by some experts that sperm whales, especially adults, were essentially immune to the threat of predation. But while it might be imagined that no predator could challenge the leviathan sperm whale, there is in fact one animal that is capable of it. The killer whale, or orca, is both a highly intelligent hunter and massive enough to tackle adult sperm whales. According to some, the name "killer whale" is an adaptation of the Spanish *asesina ballenas*, literally "whale killer." They earned this label from the accounts of Spanish fishers and whalers who saw the orcas hunting larger whale species. Given our admiration for orcas and our squeamishness about their name, the term "killer whale" is increasingly going out of favor. "Orca" is a contraction of their scientific name, *Orcinus orca*, although this, in turn, has its own negative connotations, at least for etymology lovers, as "Orcinus" can be translated as "from the kingdom of the dead." So much for orca PR.

The fact remains that these are among the most innovative, intelligent, and ruthless hunters on Earth.

Accounts exist of orcas attacking sperm whales, but few are as compelling or as harrowing as that provided by Robert Pitman and his colleagues at the US National Marine Fisheries Service from 1997. They described witnessing an attack by up to thirty-five orcas on a group of nine sperm whales off the coast of California that began sometime in the early morning and continued over several hours. The scientists who saw it, from aboard a US research vessel, said the sperm whales gathered together into what is known as the *marguerite formation*, a typical sperm whale response to a threat. *Marguerite* is the French name for a daisy, and it is an apt description, because the sperm whales turn their heads to the center of the formation while their bodies radiate outward like the petals of the flower. Young, vulnerable individuals gravitate to the center of the daisy, out of harm's way. It's even been known for other, smaller cetaceans, such as pilot whales, to seek this sanctuary. With the adults' heads pointing inward, they can brandish their most potent weapon, their tails, at the attacker. The strategy has much in common with the communal defense of animals such as musk ox, or even the infantry squares formed by soldiers long ago. But even the most coordinated of positions can be overrun and, on this occasion, the sperm whales were heavily outnumbered. The orcas' strategy was cautious, based on attrition: they gradually weakened their quarry while minimizing the risk of injury to themselves. Taking turns, the hunters would wound and withdraw, as Pitman put it. It was working—fresh blood streamed from the victims each time the orcas moved among them, and a slick of whale oil collected around the attack zone.

With blood in the water, the orcas intensified their efforts. The wounds the sperm whales suffered became ever more severe. Pitman described many with skin and blubber torn away in great sheets, and one whale with its intestines visible. Sadly for the sperm whales, it was all too clear that the end was in sight. At eleven o'clock, four hours after the orcas first struck, they finally succeeded in breaking the sperm whale's defensive formation, exposing the exhausted victims to further assault. The end came in a frenzied attack by a huge male orca, who slammed into the unprotected flank of one of the helplessly adrift sperm whales. Seizing hold of his victim, the male shook his prey like a terrier with a rat. The attack complete, the orcas could enjoy the spoils, and over the next hour they feasted on the carcass. The destiny of the other eight sperm whales was unclear. Perhaps they were able to escape, or perhaps the severity of their wounds meant that their fate, too, was sealed.

Despite the savage events that unfolded that day, successful attacks by orcas on sperm whales are actually extremely rare. Probably a decisive factor on that occasion was the extent to which the sperm whales were outnumbered by their foes. Scar patterns on the bodies of sperm whales do provide evidence of violence by orcas, however. In one census, almost two-thirds of the sperm whales observed carried bite marks from orcas. Yet Pitman's description of events remains one of only a handful of reliable eyewitness accounts. Although orcas may target sperm whale social groups with calves, the whales are usually able to rebuff the attacks. The presence of a male sperm whale in the vicinity is sometimes enough to stop the orcas from acting upon their aggressive instincts—bull whales are a third larger than their female counterparts, dwarfing their persecutors. Nevertheless, even bulls become cautious when they detect the presence of orcas in the area. Unlikely though it would be for an orca to take on a fully grown male sperm whale, the response of the males harks back to experiences earlier in their lives, when orcas did pose a threat. At the sound of orcas, these gigantic sperm whales surface, perhaps to refuel their oxygen supplies as a precaution; if they had to, they could dive down beyond the range of the whale killers and escape, and that breath would buy them time. Fleeing isn't their first instinct, however. They remain watchful, drawing closer to others of their own kind to provide an intimidating bulwark against the danger.

Pods of females and their calves are the most vulnerable to orcas. The very youngest cannot dive deep enough to elude them, and that constrains their mothers, forcing them to remain at the surface. The marguerite formation then becomes the main method of protecting the susceptible calves. It's not the only defense—there is a strong sense of altruism in the whales' defensive strategy that goes beyond the simple desire of mothers to protect calves. During the brutal attack described above, the observers related the support the sperm whales provided to one another, even while the orcas actively tried to break the formation. When on occasion the orcas pulled one of the sperm whales out of the marguerite, the isolated animal faced a terrifying intensification of attacks. But when this happened, one or two other whales would break from the marguerite to escort their lone confederate back into the defensive formation. Doing so meant the orcas would turn their attention to these rescuers, and make them the focus for savage attacks, yet it was a price they were willing to pay.

A few days later, Pitman and his colleagues witnessed another attack by orcas on a sperm whale social group. Their description of events provides a vivid and astonishing picture of the extent to which sperm whales may cooperate in defense. This time, while watching a social group of five sperm whales at the surface, they noticed more whale activity in the area. There was another group of sperm whales, including a calf, around half a mile from the first. Half a mile again beyond them, a pod of five orcas was heading toward the second group. Perhaps detecting the approaching danger, the second group submerged briefly. Why exactly they did this was unclear, but one possibility is that during their brief dive, the whales in the threatened group had sent out some kind of an alarm. Whether it was a distress call or simply the realization that there were orcas in the vicinity, the more distant pod of sperm whales changed course. Even as the two groups of sperm whales met, still more arrived, apparently heeding the call, perhaps returning from feeding dives, until the original targeted group was now fifteen strong. This show of solidarity alone did not entirely discourage the orcas. An adult female orca closed in on the group and moved among them, and a skein of oil forming at the surface suggested that the orca was at work, inflicting wounds. The sperm whales' commotion and agitation was apparent, not only to the observers on the research ship, but also to other sperm whales in the vicinity. The scientists described what must have been a mind-boggling sight: other groups of sperm whales charging in from all directions, including one that apparently came from a distance of more than 4 miles, all traveling so fast that bow waves formed around their heads. Eventually, a huge aggregation of some fifty sperm whales converged on the site. Embattled sperm whales only adopt the marguerite formation when they are in small groups; with these large numbers, their strategy was rather different. They formed into a single, tightly packed, cohesive group, all facing in the same direction—toward the orcas. At this, the orcas, now perilously outnumbered, withdrew and left the scene. With the immediate danger gone, the sperm whales fragmented into their many constituent groups and dispersed.

The recruitment of dozens of sperm whales from a vast swath of the ocean to defend against a common enemy is something that invites comparison with some of the best aspects of humanity. There are many times throughout human history when families and nations have stood alongside one another, united in the face of danger. But if sperm whale altruism puts us in mind of our finest hour, we might also bear in mind a rather less appealing flip side to our own behavior. Noting the tendency of sperm whales to respond to the distress of a group member, whalers quickly learned that if they were to injure a single whale, other sperm whales in the vicinity would congregate at the site. The injured whale thus became the bait in a terrible trap, and the whales' altruistic naivete their undoing.

Sperm whales have been a focus of man's attention for hundreds of years. First we studied them so that we might exploit them during the whaling era; more recently we have developed a deeper understanding of the species through science rather than commerce. Though orcas did not entirely escape the attentions of whalers, they were never exploited to the extent that sperm whales were. One reason is that they are both smaller and have far less blubber, making them less valuable as a catch. As with the sperm whale, though, recent studies of orcas have allowed unparalleled insight to their behavior and social structures.

COMPLEX KILLERS

As soon as we start talking about orcas, we hit a problem. We might be talking about one species, or we might mean several subspecies that we've bundled together under a single banner. To complicate matters, there are coexisting definitions of exactly what a species is. In short, it's a bit of a mess. Happily, this is a book about behavior, so we can duck some of the more esoteric controversies and leave those who classify organisms to fight it out. What can be said is that there are populations of these animals all over the world's oceans and they exhibit a staggeringly diverse range of behaviors. We sometimes refer to each of these as an *ecotype*, a type of orca that occupies its own particular ecological niche—and diet is one of the most obvious points of differentiation. To give you an idea of the diversity, there's one ecotype in the Antarctic that focuses on eating penguins, while living side by side with another ecotype that eschews penguins in favor of seals, and a third that rises to the challenge of hunting other whales. There are also ecotypes that rely on fish, often confining themselves to a narrow range of species. For instance, one type fixates on cod, another almost exclusively consumes herring, and still another considers rays and sharks to be the only worthy thing on the menu. Along the Pacific Northwest coast of the United States and beyond into Canadian waters, two ecotypes coexist. The so-called resident orcas of that region feed exclusively on fish, while their counterparts, known as "transients," are mammal hunters. This isn't merely a preference: in less enlightened times, during which mammal-eating transients were caught for entertainment purposes at marine parks, their refusal to eat the fish they were provided with in captivity was so resolute that they starved to death.

Diet on its own doesn't define a species, or even a subspecies, but the rarefied intelligence of orcas, in combination with the tendency of ecotypes to specialize, leads to a whole host of fascinating behaviors. In the seas off New Zealand, orcas rely heavily on rays for food. These close relatives of sharks have flattened bodies that equip them for a life of foraging on the ocean floor. Smartly, the orcas use a mysterious facet of the ray's biology as a weapon against them. A deft flip turns the ray upside down and induces it to go into a kind of defenseless trance, known as *tonic immobility*. At other times, the orcas work in pairs— one grabs the ray's tail and pulls it from its cover on the seabed, and its partner then delivers a killing bite to the head. Having dispatched a ray, they will share it out among a group like a fish pizza. Incidentally, orcas elsewhere use the same entrancing trick, most notably in an incident that occurred off the Farallon Islands near San Francisco. This is a gathering place for some of the world's largest great white sharks, some of which approach an orca in size. Yet according to an eyewitness account, an orca managed to invert one of the sharks and send its fearsome prey into dreamland before killing it—a handy tactic, and one that speaks volumes about the abilities and skills of these intelligent animals.

In the North Atlantic, orcas work together to separate manageable groups of herring from the massive schools of the fish that overwinter in deep water. Next, the orcas corral their prey toward the surface. The hunters encircle the group in a deadly carousel fashion while sending curtains of bubbles from their blowholes and flashing their white bellies toward the fish. Startled by the displays and relentlessly harried by the ring of orcas, the herring gather into a tightly packed shoal, allowing the mammals to apply the coup de grâce. With a dexterous, whip-like flick of their tails, the orcas send powerful concussive pressure waves against their prey, stunning them. It only remains for the orcas to pick off the inert bodies of their victims. It's not unknown for humpback whales to gate-crash the party once the orcas have done all the work. A single well-timed ascent and lunge and the orcas' efforts are all in vain, as a school of carefully herded fish is swallowed entire by the humpback's huge gape. Though other orca ecotypes may occasionally target larger whales, the humpback has little to fear from this pack of fish eaters.

Mammal-hunting ecotypes face different challenges. Specializing in intelligent prey such as seals and whales has demanded that orcas develop sophisticated predatory strategies. The results are dramatic and have captured the attention of wildlife filmmakers the world over. In Patagonia, orcas time their arrival at sea lion rookeries to coincide with the weaning of pups. Rather than wait to pick off the naive pups as they venture from their birthing grounds, the orcas make use of attack channels to get close to shore. An immense burst of speed drives one of the hunters up onto the beach, where it exploits the element of surprise to snatch an unwary sea lion. Meanwhile in Antarctica, teamwork and an exquisite understanding of physics allow orcas to knock seals from floating ice floes. The orcas coordinate in a group charge toward the ice and generate a wave that either washes the seal from its refuge or overturns the ice sheet entirely and dumps its occupant into their welcoming embrace.

Collaboration is especially important when orcas tackle their close relatives, baleen whales. The size of many adult baleen whales makes them a near-impossible challenge, but calves are vulnerable. The orcas' goal is to separate their target from its mother, which requires unstinting effort. They ram and bite the unfortunate whales and push their bodies between mother and calf to isolate the youngster. If they manage this, the orcas launch themselves across the weakening calf's back, pushing it beneath the waves and cutting off its oxygen. It's a difficult spectacle to watch, and you'd be hard-hearted not to feel for the victims in their plight. But it does demonstrate the orcas' brainpower. The extent to which they unite to outwit their prey is the equal of animals like chimpanzees, with whom they share a high level of intellect. What it also suggests is that orcas have culture, the ability to learn and to accumulate knowledge through the generations, shaping the development of phenomenally successful hunting strategies.

The phrase "you are what you eat" was never truer than with orcas; the extent of their diet specialization and the cultural development of ecotypes charts a course that leads to ever greater distinction among them. Each ecotype tends to associate and breed only with others of its own kind, and each has its own specific pattern of vocalizations, akin to a language, and often its own color pattern. For instance, though the transient and resident orcas of the northeastern Pacific share the same waters, they scarcely interact; indeed, they appear to avoid one another. Beyond these differences in appearance, diet, and dialect, there are also striking differences in social behavior.

The mammal-eating transients typically live in small groups, or pods, of around three whales—an adult female and one or two of her offspring. These pods will sometimes mingle in larger groups, but these are temporary meetups, and each pod will eventually go its own way. Though their groups are small, they do hunt cooperatively, with deadly efficiency. Each individual takes a different role, one leading the attack, others flanking it in support. When a seal is cornered, one orca acts as a blocker, sitting deeper than the prey animal, preventing its escape, while others take turns to hit at the unfortunate prey with their tails or pectoral flukes. When hunting fast-swimming porpoises, they tag-team in pursuit, alternating in the role of chaser, so that as one orca tires, another can take the lead, until, inevitably, the porpoise becomes exhausted. Regardless of the prey, these orcas are very successful hunters, perhaps the most successful of any hunters that specialize on mammals as prey. Robin Baird and Larry Dill, researchers from Simon Fraser University in British Columbia, described observing 138 attacks, of which all but 2 were successful. Once they launch an attack, it seems a question of when, rather than if, the prey will succumb.

While they're searching, transients often cease vocalizing, keeping "radio silence" so as not to give advanced warning of their approach. Once they're locked on to a target, however, communication opens up again as they coordinate their efforts. The single-mindedness of the pursuit stands in sharp contrast to their behavior once the hunting is out of the way—then they become seemingly carefree, rolling around one another, slapping fins, or breaching. Having made the kill, the orcas divide it up among the hunting party. Baird and Dill reported how one orca would approach another with a seal in its mouth. Both would then take hold of the carcass and pull it in two like a meaty Christmas cracker.

The resident orcas that live alongside the transients for part of their range are much more gregarious, living in pods that might number in the teens. Partly, this reflects the different demands of hunting fish. Once a shoal of their prey has been located, the resident orcas don't need to coordinate their hunting efforts as closely as their mammal-hunting counterparts do. Just as importantly, an increase in the number of hunters chasing fish at a particular site doesn't interfere with their individual success. Their social groups are built on the typical mammal foundation—the matriline: all the members of the pod are the descendants of a single female, with as many as four generations coexisting at once in one of the most enduring family relationships in the natural world. They're held together by the matriarch in much the same way as in elephant society. To muster the family, the elderly female sometimes calls them to attention by slapping her tail on the surface of the water, a sound that is conveyed far and wide by the excellent transmitting properties of water. She acts as a repository of crucial information, using her experience to lead the pod to good foraging grounds; she even catches salmon to hand out like party favors among her family.

Southern resident orcas are among the best and longest studied of any marine mammal. Work over more than forty years has provided fascinating insights into these charismatic creatures. Collectively, the southern residents are a clan made up of three different pods that spend much of the year in the Salish Sea, an inshore marine area protected by Vancouver Island. We know from this research that orcas are among the longest lived of any mammals—Granny, a member of the J pod, was estimated to be over one hundred years old when she died. The data also provides answers to intriguing questions in biology. For instance, orcas are one of the very few animals that go through menopause. Most animals, once they reach adulthood, live only so long as they remain fertile; yet people and orcas enjoy a lengthy post-reproductive period. A female orca seldom gives birth beyond the age of forty, though she might live for decades longer. This is something that seems so familiar and natural to us that we don't question it, but biology isn't sentimental. Why are these aged orcas hanging around when they can no longer breed? The answer lies in the fact that promoting the success of those who share your genes is an excellent strategy in evolutionary terms. Orcas, as we have seen, live in close-knit family groups, exactly the kind of circumstances that might promote this kind of behavior. What's more, both male and female offspring remain with the pod. The adult males might abscond for a brief period from time to time to enjoy a liaison with a female from a different pod, but they always come back.

As female orcas age, calves of younger females in the pod increasingly outcompete their calves for food. Perhaps the more vigorous, youthful mothers give their offspring a head start in life in some way. Ultimately, the effect of this competition is that the older females' calves are less likely to thrive and survive, so there's less benefit to an older female reproducing. But perhaps the most startling finding is the beneficial effect that these elderly matriarchs have on the rest of the pod. We can see the full extent of just how valuable it is to have older females in the pod by studying what happens in the aftermath of their death. The mortality risk of adult female offspring is five times greater in the year following the death of their mother than if she were still living. For adult sons, the mortality risk is a truly incredible fourteen times greater. Why the difference? One suggestion is that, because male orcas mate outside their pod, they spread their genes around, and their offspring will not compete directly with their mother's natal pod. This implies that the mothers are biased toward their sons and help them disproportionately, to the point that the young males come to rely on this help and are bereft without it. Whatever the truth of this matter, it's certainly the case that mother orcas play a hugely important role in the lives of their offspring.

The ties within orca society are both strong and, to a human observer, compelling. Cooperating to hunt and later sharing the catch both reinforce social bonds. There are many examples of social animals sharing a kill—a pride of lions settling down to feast on a wildebeest is one that springs to mind. What is far less common is for animals to share small prey that can be caught and consumed by a single individual. Resident orcas are fish specialists, but their favorites are salmon, and especially chinook salmon. Despite this, they regularly share their prey with other pod members, breaking up the fish before passing it around. This behavior may occur between any pod members, but it is most common for adult females to provide food to their own offspring in this way. Moreover, there are reports of orcas supporting disabled pod members with gifts of food. Often such orcas have missing fins, and it's difficult to say whether they were born that way or lost them in an accident. Missing a fin makes it a challenge for such animals to capture prey because they are slower and less nimble. Rather than leave these unfortunate individuals to feed for themselves, however, other pod members take it upon themselves to provide them with food. In this apparent tenderness, we see the counterbalance to the image of orcas as cold and ruthless hunters.

FLIPPERY CUSTOMERS

Of all the animals I've encountered in the wild, none has given me a greater sense of being appraised than bottlenose dolphins. My most memorable encounter was in the Azores, on the same trip that I met sperm whales. One moment I was finning along, scanning for life in the glassy depths; the next, I'd been joined by a pod of a dozen dolphins, surrounding me and filling the water with their squeaking, rattling, whistling conversation. I'd love to know what they were saying about me, unless it was "Who brought the fat guy?" At one point, they switched from a loose configuration to a tightly bound unit and dove a few feet below me. There they stopped and, turning upright in the water, lingered, side by side and motionless, almost in a line, watching me. Then, perhaps unimpressed, they broke ranks, and, after circling once more, disappeared out of sight. Bottlenoses, along with their larger dolphin cousins, the orcas, and the still-larger sperm whales, are, of course, recognized for their considerable intelligence. But my impression in that moment went beyond that. I felt I was being assessed by profoundly sentient, conscious animals.

The question of animal consciousness is hotly debated. It's also one that can't be fully satisfied with our current scientific capabilities, so it becomes a matter for philosophy. We can see brains, we can even interrogate patterns of brain activity, but we cannot yet see what resides within: the mind of an animal. We can set tests for animals, such as their ability to recognize themselves in a mirror. Dolphins pass this test, which implies that, at the very least, they're self-aware. Far beyond that, we can't go. Is a dolphin conscious? Can it think about thinking? There's no universally accepted answer. My instinctual perception of being in the presence of animal consciousness has no scientific validity, for all that it was a powerful sensation. There's no question, however, that these are supremely intelligent creatures.

The quality of intelligence is one that is most often seen to its highest extent in social animals, and dolphins have one of the most complex societies of all animals. Theirs is an endlessly shifting mosaic of associations, underpinned by sophisticated recognition abilities that allow them to maintain enduring relationships with dozens of others. They travel in small groups that frequently form and break up as individuals either join or leave to seek out other companions. Males, in particular, establish bonds between themselves at a young age that can last for a lifetime. As they approach adulthood, coalitions become central to the males' strategy, especially in their aggressive sexual behavior, since males may attempt to coerce potential partners. On occasion, several male coalitions may unite to form a large gang, which can then either challenge larger groups of females, or attempt to take females from under the control of other males. By contrast, females have a wider circle of associates among their own sex, but don't share the same intensity in these relationships as the males. This is partly due to estrus patterns and motherhood, both of which influence a female's social instincts. Nonetheless, females are known to face the threat of male coalitions by joining forces on occasion, gaining protection both for themselves and for their offspring through numbers.

Bottlenose dolphins interact in a variety of fascinating ways that go much further than simply hanging out together. For instance, close associates synchronize their movements, mirroring each other as they rise, surface, and dive. This might stem from their early experiences as calves, when they fall into step in just this way with their mother, flanking her and following her every move. Perhaps it's a natural progression to behave like this with a social partner. They sometimes augment their parallel swimming with a kind of dolphin hand-holding, touching each other's pectoral flippers as they go. When there's a falling-out among dolphins, they even use their flippers to stroke each other, restoring the damaged relationship.

As well as physical signs of affiliation, bottlenose dolphins also have a fascinating and complex language and maintain a running commentary, especially when they're confronted by something new and exciting in their environment, such as a portly biologist swimming in the Azores. We are gradually beginning to gain some understanding of the dolphins' language, and in particular what we call the signature whistle. This is an identifier, a specific kind of sound made by each individual dolphin, akin to the sperm whale's coda. Like human babies, they start making a noise pretty much from the outset, and they're excellent mimics, experimenting with the sounds they hear around them. It takes time to settle on a signature whistle, though—often between a year and two years; when they do, it's strongly influenced by the adults they grow up with. It's not necessarily the vocalizations of those dolphins to whom the calf is closest that determine its own call. They pick up a mix of different influences to shape their own whistle, and they may be drawn to exotic elements of the vocalizations from occasional visitors. Once they've fixed on their sonic identity, they use it in their meetings with other dolphins. Identifying yourself in the context of an encounter is one thing, but dolphins take this further—they copy the signature whistles of other dolphins as they meet. While they're far from the only animal to be able to recognize individuals according to their voices, they are the only animal that we know of that addresses others in the wild using a specific label, a label that we would call a name.

The vocabulary of dolphins extends far beyond name-calling to include a whole host of different sounds that they make according to circumstances. There are distinct calls to draw group members together, others that they make during play, and still more to indicate distress, anger, and aggression. Meanwhile, groups of dolphins share detailed information with their group about food patches as they forage, potentially directing others to good spots. There's a squeal of delight when they catch a troublesome fish; there's even a chiding note that angry mothers produce if a calf fails to come to them when called. Calves are easy prey for sharks, as well as potentially being targets for the aggression of other dolphins, so it's little wonder that the mothers are worried when they stray. A mother may discipline the errant calf further by pressing her beak against the calf's side and angrily buzzing, or, in an extreme case, by holding her calf against the seabed for a few moments. After a telling off, the calf may seek to pacify its mother by stroking her head with its pectoral flipper. Chatterboxes though they are, dolphins also know that there are times to be quiet—silence descends on the group when sharks are on the prowl.

Dolphin vocalizations are both extensive and complicated. Rendering the different sounds into our language, using words such as squeak, whistle, bray, squawk, and pop, fails to communicate the complexity and subtle variations in sound that dolphins are capable of. We face a basic barrier to understanding. In research, the verbal descriptions of dolphin sounds can be backed up by images, particularly from spectrographs, which show the calls as a graph of sound frequency and amplitude against time, but these can be difficult to interpret. Recently, a technique using an instrument called a

CymaScope has been used to capture the richness of dolphin communication. The CymaScope translates individual dolphin calls into images called CymaGlyphs that are based on the vibration patterns they produce in water. It's thought that echolocating dolphins use the sound waves that bounce back to them from objects in their environment to effectively construct images of those objects, so that, in a sense, they use sound to see. CymaGlyphs potentially allow us to visualize their calls in a similar way. It's early days, but the potential to reach further into a language of another species is tantalizing.

As befits an animal with such a well-developed brain, dolphins are both exceptional innovators and adept learners. Accordingly, they have developed some of nature's most extraordinary foraging strategies. In the Gulf of Mexico, dolphins will entrap fish by swimming in a tight circle, beating their tail flukes rapidly against the muddy substrate to produce an encircling barrier of silty water. Perhaps perceiving themselves to be confined within this wall, the fish try to jump their way clear. This is exactly what the dolphins want, and they expertly catch the fish in their open jaws. Meanwhile, in tidal creeks, other bottlenose dolphins perform strand feeding, driving fish up against the sloping banks of the creek and pushing both themselves and the fish out onto the mud, where the dolphins can pick off the marooned fish with ease. And in Shark Bay, in Western Australia, there's a select band of dolphins who collect sponges from the seabed; they drape these over their lower jaws to protect themselves when they're trying to winkle a recalcitrant fish from a hiding place beneath a rough substrate.

Making the transition from having your food provided for you, in the form of milk, to catching your own lightning-quick, super-slippery fish is no easy matter. There are lots of different approaches, and dolphins can potentially learn from any of the elders within the population. But the hunting strategy of a mother is the strongest influence on the strategy that her calf comes to rely on in later life. The evidence points to the dolphins passing these behaviors from generation to generation. For instance, the sponging behavior is thought to have originated with the "eureka moment" of a single female dolphin almost two centuries ago. Intriguingly, once a dolphin develops a particular strategy, it tends to shape not only how it forages, but also who it associates with. Sponging dolphins primarily interact with other spongers in a kind of specialist's club. Though picking up these skills might be simply down to a calf watching and imitating its mother, there's some evidence to suggest that some dolphin mothers deliberately teach their calves their hard-won life skills. When they're accompanied by junior on the hunt, mothers seem to direct the youngster and to undertake longer, more drawn-out chases than they would do on their own, possibly as a means of providing experience.

DO HUMPBACKS HAVE THE HUMP?

Fifty years ago, at the height of the hippie era, a record was released that captivated the world. Unusually, it was recorded—in a sense—by an animal. *Songs of the Humpback Whale* became a best-selling album, its haunting, emblematic sounds encouraging a rethink in attitudes toward these creatures. It helped popularize and gather support for the fledgling Save the Whales movement, ultimately leading to the moratorium on whaling. Recordings of humpback songs were even carried on the Voyager spacecraft as part of its message to whatever alien civilization might intercept it.

I had this in mind when, a few years ago, I was perched on the side of a boat off Tonga, waiting to jump into the sea, where, a little way away, a humpback had launched into song. I was anticipating a revelatory experience: the mysterious, even mystical song would be revealed to me in a moment of surpassing beauty. I jumped in, camera already recording, so that I could play back the sound and revel in its ethereal majesty in perpetuity. There he was, magnificently poised, his head down, his gargantuan body at an oblique angle, his great pectoral flukes held out to either side like an opera star in the full throes of musical drama. Just one small thing was wrong—his song was awful. I'm not being overly critical—I don't mean he fluffed a note. For all I know, listening humpbacks in a 50-mile radius were entranced by its soulfulness. But he sounded like a pig coming round from a hangover. He was among a select and tiny group of animals on Earth who are worse at singing than me.

During my visit to Tonga, I had the good fortune to spend time in the water with twenty or so different humpbacks on winter break from their rich feeding ground down in Antarctica. Aside from mothers in close attendance with their calves, and one instance where two huge males were in hot pursuit of a female, they were individuals on their own. There was no great surprise in this; it confirmed what I'd learned years before as an undergrad: whales are broadly divided into two groups, the toothed whales, such as sperm whales and dolphins, and the baleen whales, such as humpbacks, blue whales, right whales, and so on. The received wisdom was that the two groups could be neatly divided according to their sociability. Toothed whales are gregarious, apparently, and baleen whales aren't. Humpbacks and their kindred live and feed on their own and are solitary by nature. They don't need the protection offered by groups; nor do they struggle to find their prey—in fact, getting too close to one another while feeding could interfere with their efforts. Not the most gregarious animals, we might assume, and not especially well suited to a book about sociality.

But there's a big difference between assuming something and knowing it. The whole point of science is to rigorously challenge ideas, and in recent times evidence has emerged to change our perspective on humpbacks and their social world. One of the defining characteristics of an animal group is that its members are in sensory contact with each other. That's not a rule that works well for whales, who can exploit the sound-transmitting properties of water to communicate over enormous distances of hundreds of miles. Humpbacks can stay in touch and maintain relationships remotely. From some stunning findings that emerged in the late 1990s, we know they're listening and taking notice of each other. While humpbacks are distributed widely across the seas of the world, the males within each ocean basin tend to sing more or less the same song. It isn't fixed, but changes gradually over time: the males within a given area adapt their ballads according to what they hear and whatever's current among the tunes being belted out by the other males. Sometimes, the song may be completely revamped. Beginning in 1997, the humpbacks on the east coast of Australia began singing a new melody, one that they'd picked up from their counterparts on the west coast some 2,500 miles away. Whether they'd learned it from a few males that had strayed over to the Tasman Sea, or instead overheard it while migrating or feeding in the Antarctic, isn't known. But by 1998, all the eastern humpbacks were crooning the west coasters' song. It was a cultural revolution in musical form, and it was one that was shaped exclusively through social influence. We now know that these revolutions do periodically occur. The whales' music gradually gets more and more complex as individual males add embellishments to their tunes, possibly as a means to stand out in a crowd of singers. Their fellow males then respond by building in some of these elements, with the effect that, over the population, the song grows and develops. Then, over a relatively short period of time, there's an overhaul, a revolution, and the whales go back to basics with a different and simpler song, beginning the process again.

Though the appreciation of whales as social song learners was a game changer in its own right, it's far from the only evidence we have to refute the idea of humpbacks as committed loners. Off the coasts of North America, humpbacks cooperate in small groups to trap schools of fish as prey. To pull off a fish heist, the whales dive deep and then begin to release a stream of air from their blowholes. Next, working together and coordinating their efforts through vocalizations, they spiral upward, producing a columnar curtain of bubbles that corrals and entraps the fish within. Finally, as they approach the surface, one of the whales gives the signal to attack and they all lunge at the fish, engulfing them in their cavernous jaws. Occasionally, the whales add to their repertoire. In 1980, a single humpback in the Gulf of Maine was seen smashing at the water's surface with its tail. This unusual behavior turned out to be an innovation, possibly developed by this one inventive whale for feeding on sand lance, a small fish that gathers in the locale to breed. In the years since then, lobtailing, as it has become known, has spread in the population. Like the songs of the males, it's a learned behavior, and one that has diffused through the whale network, from one to many individuals, based on close associations.

Our conception of humpbacks is as gentle giants. In Tonga, where the whales gather to breed and give birth, I came across a young female that was motionless at the water's surface. I tentatively approached, unsure what I'd find—was she sick? Her eyes were closed, but as I swam nearer, she opened them and regarded me solemnly. After a few seconds of inspection, she apparently decided that I was of no consequence, and shut her eyes again. I circled her at a distance to check if she was carrying injuries, and then, finding nothing of concern, returned to the boat. I think she must simply have been sleeping, because, an hour or so later, she was enjoying a game with a leafy branch she'd found floating on the water, flicking it with her pectoral flukes and pulling it around in her mouth, like a kitten with a toy. Just about every humpback I came into contact with responded to me in the same passive way, neither aggressive nor fearful. Based on that, the idea of them as laid-back leviathans seems well founded, but it's not entirely accurate. During the breeding season, frisky males compete, sometimes violently, for the affections of a female. The most extensive demonstration of humpbacks' power is, however, directed toward their only significant marine foe, the orca. Although measuring over 40 feet in length, an adult humpback has little to fear from killer whales, but for calves the picture is more concerning. Estimates suggest that as many as one in five humpback calves may be killed by orcas, while many adult humpbacks bear the scars of failed attacks, such as rake marks of orca teeth on their flukes. If humpbacks are capable of bearing a grudge, then orcas are likely to bear the brunt of it.

It's been suggested that one of the reasons that humpbacks migrate to the tropics to give birth is to avoid the depredations of orcas on their calves. While orcas are widespread throughout the world, they're far less common in warmer waters. The threat that these supreme predators pose might shape the humpbacks' migration routes as well, with females who have calves staying closer to shore, hugging the coastline. That said, humpback mothers are far from helpless, and they're capable of energetically defending their young against attack—to the extent of lifting the calf bodily from the water and positioning it on her back. The challenge grows as the number of orcas increases, and there's little that even a determined mother can do under a concerted attack from a large pod of killer whales. Orca danger may partly explain why some mother and calf pairings are accompanied by an escort, usually a male. It's only a partial explanation, because the goal of the male is most likely to breed with the female; nonetheless, he can help defend the calf, who is better protected by two adults flanking it than one. These pairs, too, will occasionally lift the infant clear of the water and move it to their backs to prevent the orcas from ramming it or bringing their deadly teeth to bear.

Perhaps a lifetime of animosity explains why humpbacks sometimes turn the instinct for defense into attack. Robert Pitman and his colleagues described a number of occasions where humpbacks actively sought out orcas as they were in the act of hunting, even when the predators heavily outnumbered them. The particular species the orcas were harrying seemed not to matter: seals, sea lions, other whales, or fellow humpbacks—the whales were drawn from considerable distances by the sounds of the orcas making their assault, and then weighed in with some lusty blows on the victim's behalf. It's a remarkable thing to do, for although orcas are somewhat smaller than adult humpbacks, they are nonetheless a dangerous adversary. The humpbacks appear to be unperturbed by this, lashing out with their most effective weapons—their tails and their huge pectoral flukes, each of which can weigh a ton. Consistent and frequent interventions like this are almost unheard of in the animal kingdom. Mobbing behavior, where prey animals turn the tables on an attacker in a show of force, is known in a few species, but it's most often carried out in defense of kin, or, at the very least, to protect other members of the same species. While it's entirely possible that the humpbacks rally to the cause on the mistaken assumption that the orcas are targeting their own kind, they continue to harass the predators long after they've discovered the identity of the prey. The benefits to that victim are, of course, enormous. The gains made by the humpbacks are less obvious; perhaps they simply enjoy avenging themselves against their greatest enemies.

Over the past half century, there has been a sea change in the way we appreciate whales and dolphins. Although our historical relationship as predators and prey is regrettably continuing in some parts of the globe, most people now have some understanding of how complex and fascinating these creatures are, and know they are imbued with an intelligence that overreaches that of almost every other species on the planet. In whale and dolphin societies, we see compelling and enduring relationships, sophisticated interactions, and strong evidence of an animal culture. While in some regards we've been slower to appreciate the intricacies of the lives of humpbacks and their like, in comparison to the toothed whales, each passing year brings new and extraordinary revelations. Though many of these animals are recovering from centuries of exploitation, for others there is still a long way to go. The problems are compounded by our insistence on treating the marine environment as if it were infinitely capable of swallowing our garbage and our pollution. It's to be hoped that we can learn to treat these intelligent animals, and those that share their marine habitat, with greater consideration.

CHAPTER 9

WAR AND PEACE

Our closest animal cousins enjoy fascinating
and complex social lives.

CLOSE COMPARISONS

There are about to be five hundred different species of primates in the world. They range from the tiny mouse lemurs that could sit comfortably in the palm of your hand to the imposing yet largely peaceful gorilla, which couldn't. Primates are the new kids on the animal block, having appeared around sixty-five million years ago, long after other major vertebrate groups. It was another twenty or thirty million years before the first monkeys made their entrance, while apes appeared more recently still.

We share a common ancestor with modern-day chimpanzees—it's a mere six million years since our two bloodlines went their own separate ways. This might sound like an extremely long time, especially in an age when we get tetchy about waiting so much as fifteen minutes for a meal at a restaurant, but in evolutionary terms, six million years is a flash in the pan. While we might consider ourselves to be far removed from our modern primate cousins, our common heritage, and the huge overlap in our DNA, means that the distinction is smaller than you'd imagine.

Perhaps that's why we see so much of our own behavior reflected in these animals. This is especially the case when we consider our tendency to seek out the company of others, to form relationships, and to live in families. The vast majority of the world's primates are, like us, social. And, like us, they initiate affiliations and rivalries, they communicate and make decisions as groups. How they cope with the challenges their society throws at them can tell us much about our own social origins. In short, they offer a kind of master key for unlocking the mysteries of our own relationships and societies.

Although being social is far from the only characteristic we share with other primates, it may just be the driver of the other standout facet that typifies us: intelligence. If you're lucky enough to have met the gaze of a primate, you might have gained a sense of this. For most people, such an encounter will only ever happen at a zoo. It can be quite a different experience when you're interacting with them up close and personal.

My own initial exposure to their cunning came in Kenya at the hands of an enterprising vervet monkey. These are attractive creatures whose black faces are fringed with white hair, while their bodies are covered in gray fur. The vervets sauntered around the outskirts of Mombasa at their leisure, miniature, expressive, hairy facsimiles of ourselves. Foolishly, when we'd gone out to watch the monkeys, my wife and I had left the window to our room slightly ajar. This presented an open invitation to one daring monkey, who expertly negotiated his way into the vacant room. He didn't, as you might imagine, turn the place over, but instead headed straight for my wife's bag, which he dexterously unzipped to steal a pack of mints. Having secured his prize, the monkey took himself to the roof of the building, from where, periodically, over the course of the next half hour, a cellophane wrapper would float down. Eventually, the empty bag followed. Though I didn't begrudge the monkey his ill-gotten gains, I was concerned that sweets weren't an ideal food for a monkey. On the plus side, his breath would have been impeccable.

A minty snack is one thing; alcohol is quite another. In less enlightened times, monkeys were caught simply by leaving booze out for them, and were then sold as pets or used to boost the draw of sideshow entertainment. Vervet monkeys are one species among many who will happily partake, and it doesn't take much for a monkey to get drunk—an inebriated monkey can be scooped up where it falls. By the time it wakes up, presumably with a tearing hangover, such a monkey would have been put in a cage and carted off to a life of ridicule, perhaps wearing giddy clothes and sitting on a barrel organ. Happily, giving monkeys a Mickey Finn is comparatively rare nowadays, but in many parts of the world raiding parties of monkeys do descend on hotel terrace bars in search of the buzz supplied by drinking the dregs of patrons' drinks.

It's tempting to smile at this, recognizing our own weaknesses in the behavior of our primate cousins, but monkey problem-drinkers take some managing. You can't bar them, as you might a boorish human, so custodians are sometimes forced into more extreme measures. When vervet monkeys do have regular access to alcohol, the way they use it is surprisingly reminiscent of human populations. Around one in six of them drink heavily and regularly, while one in twenty become problem drinkers—beginning each morning and drinking until they fall into a stupor. At the other end of the scale, around one in six are teetotalers. The remainder are content with drinking relatively small quantities from time to time. Humans didn't invent substance abuse: the precursors for it can be found in some of our closest relatives and have a basis in our shared genes.

The bigger picture is that in living alongside human populations, monkeys learn to exploit opportunities that lead them into delinquency in other ways, too. A case in point here is Fred, a chacma baboon who for years in the early part of this century, along with his troop, engaged in a Cape Town crime spree. Initially drawn to urban life by the availability of food, he was most likely emboldened by offers of snacks from well-meaning tourists. Soon Fred ceased waiting politely and took matters into his own hands. He could hustle aggressively for food—a fully grown male baboon can be intimidating—and learned how to open the doors of unlocked cars and houses. The city authorities were forced to act, fearing the loss of tourist revenue, and Fred was caught and euthanized in 2011. His autopsy revealed that he had been shot dozens of times—there were around fifty bullets and pellets lodged in his body—yet this had failed to turn him from a life of crime. Nonsocial animals are usually less likely than social ones to mingle with humans: the skills that group-living monkeys acquire within their social environment help them to adapt to, and, on occasion, join, the fringes of the human world, with all the problems that this can entail.

Sometime after my encounter with the mint burglar, I was out for a predinner stroll with my PhD supervisor, Jens Krause, at Mpala Field Station in the Kenyan bush. After weeks without rain, the vegetation for miles around had been shriveled by the relentless African sun, and a narrow strip of green lured us down to the river. There were plenty of signs of animals. Tracks and turds littered the sandy ocher soil. But aside from birds putting in a late shift before nightfall, there were few creatures to be seen. Then, as we rounded a bend in the river's course, we came across a group of vervets collected around the base of an acacia tree. They regarded us indifferently for a moment, then went back to the endless routine of grooming that plays such an important role in maintaining social bonds in many animals. Not wanting to disturb them, we paused to watch them performing the age-old rites of monkey society. All was relaxed until one of them broke the tranquility with a sudden call. It doesn't do the call justice to liken it to the rapid, repeated quacks of a hyperactive duck with a blocked nose, but that's the best I can do. In any case, the effect on the rest of the troop was electric: heads instantly swiveled toward the caller before the whole group shot up the tree.

"That's their leopard alarm call," said Jens.

"There's a leopard nearby?" I asked, in a voice perhaps shriller than I intended.

"Mmm," he replied, calm as you like.

The news that a predator was lurking, a substantial predator at that, and one with a taste for primates, was unnerving to me and to the vervets, if not to my supervisor. The monkeys' response was to zoom up the tree. A favored strategy is to seek out thinner branches that will support their weight but not that of a 130-pound leopard. Like the vervets, I had no wish to become acquainted with the inside of a leopard, and I briefly considered scrambling up a tree myself. In the end I settled for chivvying Jens back to the field station at an undignified pace.

CALL THE ALARM!

While many group-living animals give alarm calls when they detect danger, vervets are rather unusual in that they're able to be specific. They have distinct calls for each of their main predators, and each of these warnings produces a different defensive response in the group. The leopard call, as we've seen, induces the vervets to rush to the safety of the trees. But this would be a bad idea when faced with a bird of prey, which could snatch them from their perch. So the vervets have a call for these predators, too. I don't know if frogs can hiccup; if so, then the vervet's raptor alarm call is what I imagine this might sound like. When the vervets hear this croaky siren, they seek shelter in dense undergrowth, where the bird can't follow them. Vervets also face the threat of snakes, such as pythons. Again, they have a unique cry, a kind of chirrup, and the reaction of the group is to stand on their hind legs and scan the ground around them. When they locate the snake, they might either band together to mob it in a show of force, or, if they're feeling less brave, use their greater speed to outpace it. Baboons and unfamiliar humans represent other threats to vervets in parts of their range, and they have alarm calls for them (and us) too. Some experienced vervet-watchers believe that the monkeys are capable of identifying as many as thirty different predators with distinct alarm calls. While they don't make a noise for large herbivores such as wildebeests, which offer no threat to them, they do when they see domestic cattle in some areas, because this means that local herders may be around.

Knowing when to call and when to keep quiet requires judgment. Sometimes it's appropriate to raise an alarm. Sometimes it's not. Predators that rely on stealth to approach closely enough to ambush their victims can be discouraged by an alarm call. If they know they've been detected, they often abandon their hunt, because their chances of success with an alert victim are low. Other predators, such as chimpanzees, chase their quarry through the trees, though, so seeking refuge is difficult. In this case, it's better not to advertise your presence; accordingly, the monkeys tend to keep quiet when their larger, more ferocious cousins are in the neighborhood.

As a child, I used to enjoy watching Tarzan cartoons on TV. The hero, swinging from vine to vine in his jungle home, was able to tap into a network of animal communication to get the latest local news. This isn't so far-fetched. Like Tarzan, even we denizens of the modern world, separated by generations from what was once the everyday threat of predation, can recognize fear in the calls of other animals. Participants in one study correctly identified the alarm calls of mammals, reptiles, and, oddly, tree frogs. This ability is widespread in potential prey animals. Some animals take it a step further—not only can they recognize a cry of alarm given by a different species, but they can infer what caused it. This means that if their own group's lookout fails to detect an approaching threat, they have a backup in the form of a host of other watchful animals, including squirrels, forest antelopes, and birds.

Perhaps as a means to justify our own vaunted opinions of ourselves, philosophers through the ages have attempted to define traits that are uniquely human—characteristics, in other words, that can be used to separate us from other animals. Language has long been assumed to be one such. But do the "words" used by vervets for different predators count as a language? And, if so, does that mean we're not so unique after all?

The answers to these questions are not straightforward. It was long assumed that animal calls were simply expressions of their basic emotions and motivations: a howl of rage, a shriek of fear, a moan of pain. The calls of vervets go way beyond this. When we speak, our words carry a symbolic representation of our meaning; there's nothing about the word "leopard," for instance, that intrinsically identifies how we should interpret it. Leopard is an arbitrary word that we learn, like thousands of others, and understand to carry a meaning, even if we don't all respond to it in quite the same way. Vervet alarm calls don't sound like human words, and there is no reason to expect them to. Most primates lack what it takes physically—for example, enough tongue flexibility, and a sophisticated enough control of their vocal tract—to be able to shape what we might recognize as words. Yet, despite a limited vocabulary, vervet calls do seem to fulfill the basic criteria of language. This view is given further credence when we record what happens in a monkey's brain as it hears a range of different sounds. The vocalizations of their own kind light up a very different part of their brains from that which is stimulated by everyday background noises. Such sounds stimulate a monkey's temporal lobe and the limbic and paralimbic areas—the exact same areas that are switched on by speech in our own brains. The more we discover about animals, the more we learn that our supposedly unique traits are, well, not unique. We are unquestionably more complex than other animals, but the differences between us are seldom absolutes. Rather, they are just a matter of degree.

Evolutionary anthropologist Thomas Struhsaker, working over fifty years ago, was the first to describe vervet vocab. Since then, numerous scientifically rigorous tests of the monkeys' recognition of alarm calls have been made, notably by famous primatologists Dorothy Cheney and Robert Seyfarth. These studies have demonstrated that the animals can readily identify the warnings from playbacks of recorded alarm calls. Green monkeys, close relatives of vervets, share their ability to use specific calls to identify predators. In the 1600s, a small number of green monkeys were transported from their home in Africa to the West Indies, where the comparatively benign local fauna enabled them to thrive. Over three centuries later in Barbados, a team of researchers gave these monkeys a rather nasty surprise: they played them a recording of the leopard alarm call that their African cousins made. There are no wild leopards in Barbados, but the green monkeys hadn't forgotten. On the cue of the call, they ran up a tree. Perhaps the knowledge of the leopard call's meaning and the terrible sense of danger that it evoked had lain buried in the monkeys' psyche for dozens of generations. Or perhaps, instead of "leopard," it just means "run up a tree." Yet if it does mean "run up a tree," there is no evidence of this: the researchers never heard the monkeys make this call other than in response to a picture of a leopard.

Before this analysis becomes a paean to the sophistication of monkeys, I should recount an extra detail. After I left Kenya, Jens, my colleague who had reacted so nonchalantly to the threat of becoming a leopard snack, decided to see if he could trick the vervets into making their call. On the next supply run to a nearby town, he searched for something to help him mimic the predator. Nanyuki is a small place, and not overrun with shops specializing in this kind of apparatus. Nonetheless, he found something and took it back to the research station. So it was, a few hours later, that Jens—a tall, slender man—could be seen striding toward a group of vervet monkeys wearing a leopard-print dress. It worked. The lookout gave the leopard alarm call and, quite possibly, never lived down the shame. You might conclude from this that the monkeys are a bit stupid. But in the dangerous world of the African bush, it's best to take a safety-first approach—to give the warning rather than dithering while deciding whether the approaching object is a leopard or just a man with questionable style.

MONKEY BUSINESS

For social animals, alarm calls are a means of protecting other group members, often relatives, from the risk of being eaten. You'd be mistaken, however, to think such behavior is purely altruistic, given that the alarm can directly benefit the caller, too, in various ways, such as by letting a stalking predator know that it's been spotted, and thus discouraging it, or, in the worst case, by creating a sense of panicked confusion among the group that allows the caller to escape. Over and above this interest in self-preservation, being the alarm caller, the one protecting the group, can be sexy. Monkeys seem to be aware of this. Male vervets, for example, are more likely to give the alarm when they're among females than when they're with the guys. In other species, where a single male maintains exclusive access to a harem of females, being attentive to the threat of danger and giving an alarm call not only serves to protect his offspring, but might just be the price that he has to pay to persuade the females that he is worthy of their continued favors.

Deception is a fundamental part of animal communication, and there are plenty of fascinating examples of primates using it to their advantage. In monkey society, there is often a strict dominance hierarchy that determines the cut that each individual gets of the communal cake, especially in relation to food and sex. When a troop of monkeys stumbles upon a cache of food, the low-ranking members of the group have to wait their turn while the dominants eat their fill, if not for a bit of monkey business. When Portuguese voyagers began exploring the New World, among the wonders they encountered was a kind of pink-faced monkey with brown fur over most of its body, and creamy white hair on its shoulders and fringing its face. Sardonically, the Portuguese referred to these as capuchins, for their similarity to the cowled monks of the same name. It wasn't complimentary to either the monkeys or the monks, who were regarded in some quarters as being venal and pompous. The name stuck nonetheless.

Living in groups of up to forty individuals, capuchins have to be supersmart to survive, and even to get their fair share within a highly stratified community. One proof of their intelligence is how low-ranking monkey underdogs sometimes use trickery to gain food, without taking a beating from their superiors. After giving a false alarm call, which sends the higher-ups running for cover, these subordinate hoaxers dash in to grab the unattended morsel.

Tactical deception, as this kind of sneaky behavior is known, is so recognizably a facet of human behavior that it comes as something of a surprise to see it in animals. It implies sophisticated intelligence, the ability of the deceiver to plan and predict the behavior of its mark. In line with this, it appears that its use correlates with cognitive complexity, or, more simply, brainpower. Unfortunately, it's difficult to study this behavior in a rigorous scientific manner, especially in the wild. For one thing, it's rare. Primates living in a social group interact with one another over extended periods of time. Any individual who regularly attempts to deceive others in its group will rapidly get a bad reputation. Even if the behavior isn't directly punished, other group members start to pay less attention to the perpetrator. In a human context, you might think of con artists and traveling snake oil salesmen, who could only maintain a regular stream of business from the gullible by moving from one town to another.

Among vervet monkeys, any individual who regularly makes an inaccurate alarm call—or "cries wolf"—soon finds its mates ignoring its warnings. Another reason for the relative scarcity of firm evidence about deception is that it's hard to be entirely confident about an animal's motives. In the example of the food-pilfering capuchins, a low-ranking potential food thief must experience all manner of conflicting emotions as it approaches its feasting superior. In particular, it may be fearful of the reprisals that could follow from a bold approach. In this state of fear, its nerves on edge, it may be more likely to utter an alarm call more or less involuntarily.

Nonetheless, there are some fascinating examples of what seems to be deception among primates. Take, for example, the scandalous behavior of a young baboon who, on finding an adult working hard to dig a tuber out of the ground, gives a scream that would normally only be used when under attack. The scream draws the attention of the youngster's furious mother, who charges over to attack and chase the digger, leaving the trickster alone with the delicious, newly uncovered root. In the same group, an older baboon picks a fight with a smaller counterpart, only to find the odds suddenly shifting against him as a number of adults come to the defense of his target and converge on him with aggressive intent. Thinking quickly, the aggressor rises on his hind legs to look fixedly into the distance. This behavior normally indicates an imminent threat—the approach of a predator or a rival group. The posse of rescuers pick up the cue and turn toward the direction he's looking. There's nothing there, of course, but the distraction has taken the heat out of the situation, and the protagonists simply return to what they were doing before.

The great apes are masters of subterfuge. Chimpanzees, in particular, can be highly strategic in their interactions, both with one another and with us. Santino, a bad-tempered chimp at a small zoo in Sweden, would prepare for the arrival of visitors by collecting a cache of stones from around his compound and hiding them near the viewing area where the public would gather. To conceal his plans, he would cover the stones with a scattering of hay until he was ready to spring to the attack, and fling missiles at his unwary admirers. Other examples demonstrate that chimpanzees readily understand the need to project an image, or to conceal their intentions. A chimp approached by a higher-ranking member of its troop may give away its unease by assuming an expression of fear, which, for chimps, shows itself as a grin. But in one famous example, a chimp approached from behind by his rival took time to compose himself, smacking his lips together to rid himself of the fear grin before turning to his foe, wearing his game face. Chimps are also careful to conceal other emotions. When a low-ranking member of the group finds food, if it lets its excitement show, the chances are that the prize will be stolen. Sometimes, a chimp may sit on its delicious discovery, hiding both the food and its delight, affecting an air of insouciance until the others aren't looking and it can quickly snaffle its snack without interference.

Then, of course, there's Koko the gorilla, a Pinocchio among apes. Koko was raised at San Francisco Zoo and from infancy was taught sign language by her human trainers. She excelled at this, to the point where she apparently recognized and used hundreds of different symbols, enabling her to interact with her carers to a remarkable degree. After a while, Koko identified that something was missing from her life—she wanted a pet. Cuddly toys were offered, but they didn't quite cut it, so, for her twelfth birthday, she was allowed to choose a kitten from an abandoned litter. Though Koko formed a close, nurturing relationship with the kitten, which she named All Ball, her devotion to her beloved companion didn't stop her from trying to frame the poor puss. Perhaps Koko was having a bad day when she ripped the sink from the wall of her enclosure. Inevitably, Koko's trainer confronted her about the damage, at which she pointed at All Ball and signed, "Cat did it."

Deception is a useful, if risky, strategy for advancement in competitive social environments, particularly for those at a disadvantage. The flip side is the remarkable sense of fairness shown by some primates. When their expectations aren't met, this fairness is tested, and they can react badly. Almost a century ago, the splendidly named psychologist Otto Tinklepaugh set out to vex monkeys in the name of science. He showed macaques two inverted cups and allowed them to watch as he hid either a bit of lettuce or a piece of banana under one of them. He then took the monkey out of the room for a little while before allowing it to return and choose one of the cups. This is child's play to a monkey—it would rush to the cup where it had seen the food being hidden and claim its prize. The twist was when Tinklepaugh showed the macaque that he was hiding a banana, but switched it for lettuce while the monkey was out of the room. Bananas, as any keen comic-strip reader will know, are ambrosial to monkeys, whereas salad is merely humdrum. The returning macaque, anticipating a banana, hurries in to turn the cup over only to find a lettuce. It looks around the room for its lost fruity treasure until, probably with a dawning sense that it has been cheated, it shrieks its disgust at the observer and stomps off, leaving the disappointing lettuce behind.

This sense of unfairness is even more acute when these social animals realize they're getting a bad deal compared to someone else. The experiments on capuchins by Sarah Brosnan and Frans de Waal neatly showcase this. They trained capuchins to trade pebble tokens for food. Most of the time, the monkeys would hand over their token and receive a piece of cucumber, although they didn't get to see what was on offer until after they passed the pebble over. Cucumbers are acceptable to capuchins—not anything to get excited about, but they'll eat it. A grape, however, represents fine dining. The monkeys were able to see what other monkeys received in return for their pebbles. Next door gets a piece of cucumber, I get a piece of cucumber, that's fine. But next door gets a grape and I get a cucumber? That's not fine. Not fine at all. When this happened, the capuchins, feeling they'd been short-changed, reacted furiously, flinging away the cucumber they'd paid for in a petulant rage. For monkeys like these that live in relatively tolerant societies, the perception of fairness may be an essential part of the glue that binds them. The odd thing is that it was only females who reacted so furiously; the males were more accepting of the inequality. This could reflect the fact that females are the nucleus of capuchin society, while the males are preoccupied by sex and status; or it could suggest that the males are just more philosophical about life's injustices—it's impossible to know.

THE COLORFUL TROOP

It's thought that the United Kingdom was once home to monkeys before humans arrived, and when the weather was perhaps more forgiving. Lacking the umbrellas and sou'westers required for life in the British Isles, they petered out. Nowadays, junior naturalists in my home country must content themselves with a visit to a zoo if they wish to see them, and that's where I met my first baboons.

I've always had a soft spot for these infamous monkey reprobates. They, in turn, have made their feelings known to me, in no uncertain terms. The first incident took place in a safari park, when a troop of baboons sauntered over to my run-down car. I couldn't have been more pleased when one languidly hauled itself onto the hood and sat there, staring into the distance like the dog-faced figurehead of a rubbish boat. It was joined by another, which took its place on the roof. In no time, the car was overwhelmed with baboons standing or sitting on every flat surface. I was thrilled to be in the midst of their group, watching them interacting and grooming one another, getting on with the intricacies of their social life just inches from me. I was less thrilled when they calmly and professionally removed the windshield wipers. They weren't quite as calm as they pulled out the windshield wiper nozzles, because the tubing that the nozzles were attached to put up a bit of a fight, but they were matter-of-fact about it rather than frenzied. Just business as usual. Once they'd dealt with the more portable equipment on the outside of the car, they started to see how far the antenna would bend, and, more ambitiously, the license plate. That was enough for me. Seeing an impressive set of baboon teeth as one brought his to bear on the side-view mirror, I canceled my plan to leave the car to remonstrate and blew the horn instead. A couple of the more timid ones bolted; the remainder merely looked at me with mild annoyance. I upped the stakes and started the engine. That worked. They beat a reluctant retreat to the side of the road, where they trained their orange eyes on me with an expression of defiance as I drove off. No matter, they seemed to say, another sucker will be along in a minute.

The next time I saw baboons was on their home turf, in Africa. I was teaching on a field course from the University of Leeds and driving across the Kenyan savannah in a minibus full of students. A large male swaggered over toward the bus and, as we stopped to admire him, sat on the ground, facing us with an air of command. Once he had our attention, he spread his legs and popped out his startlingly pink penis. I'm given to understand that this kind of penile display means "Stay away from my women." I think he overestimated my interest in his women, but I couldn't help but admire the self-assurance of his approach. Although facing down a bus full of students with such a singular gesture might seem a little unusual, a male baboon will never reach the apex of his society without a little braggadocio. His is an unsentimental world of threats and challenges amid a host of rivals. To reach the top and stay there, he has to be

ready to confront anything that might undermine his position (although taking on a minibus is probably a step too far).

Rather lazily, I'm using "baboon" as a general term here. In fact,
there are five different species distributed across Africa. Four of these,
sometimes referred to as savannah baboons, are so similar in their
appearance, as well as in their behavior and ecology, that it's sometimes suggested they're all just variants of a single species. The debate
rages long and fierce on this matter. Rather than get bogged down in
the detail, it might be wise to sidestep the taxonomy and concentrate
on their behavior. Savannah baboons are spread from Senegal and
Guinea in the west through the equatorial belt of the continent to
Somalia in the east and right down to the Cape. Although they'll
take to the trees when they need to, they spend their days roaming
the open savannah, foraging omnivorously on grains, berries, roots,
insects, and any small animals they can catch.

While many monkeys live in small groups—a family unit or
a single male with a few adult females in tow—baboon society is
large. Their groups can be over a hundred strong, a mix of males
and females and their young. Yet though they live in mixed societies, the experiences of the sexes are very different. The male world
is one of aggression and sex; that of the female is about the bonds
of kinship and intrigue. Males are built for fighting, weighing up
to 90 pounds—twice the weight of the females—and armed with
vicious-looking teeth. There might be as few as three or as many as
fifteen adult males in an established troop, each contending for the
exalted rank of alpha. There's nothing subtle about the competition.
They issue shatteringly loud "wahoo" calls as challenges to other
males. The most powerful among the males produces the loudest and
most intimidating calls along with a huge amount of showy bluster.
This rivalry can descend into chases, with a group of males bouncing
around their environment wahooing at each other. Only the strongest can maintain the intense physical effort needed for such an energetic and noisy display—hence, it is those males in peak condition
who win out in these signaling bouts.

If the verbal sparring fails to decide matters, the situation can boil over into a fight. Powerful as the monkeys are, fights among them can lead to serious injuries, or even death. Their imposing canine teeth, sharpened in preparation by grinding their teeth together, inflict terrible wounds. Attacks are aimed at their opponent's face, and fended off with forearms, so it's not unusual to see torn faces and damaged limbs in the aftermath. The costs of conflict can be substantial for the victor as well as the vanquished—wounds may not heal, and predators may take advantage of weakened combatants. It's for these reasons that the males put so much effort into display and bluster—it lessens the chance of an actual fight. Those that lose a fight may be displaced in the hierarchy, sometimes sinking into what looks from the outside like a kind of depression, withdrawing into themselves and moving to the periphery of the group. A fall in status can act as an open invitation to lower-ranking males to take advantage of a rival's weakness to boost their own position, so the loser must keep looking over his shoulder. Occasionally, if the turmoil becomes too much, the loser might even leave the group altogether.

The rewards for the male who reaches the pinnacle of baboon society come in the form of sex and food. The alpha gets the lion's share of matings and thus fathers a disproportionate number of young. He safeguards these spoils jealously, forming close consortships with females at the time of their peak fertility to ward off rivals. If the term "consortship" seems to imply attentive and solicitous care, however, think again—dominant males not only try to bully other males out of the running, they aggressively intimidate and coerce the females, too. The females aren't passive about this, of course. Although mating with the alpha makes for a good strategy, they like to keep their options open. They have to be sneaky, though—a straying female potentially faces reprisals. Detaching herself from the coterie of females within the alpha's orbit, she edges into a thicket for a hidden assignation with another male. Whether she knows it or not, this is a smart move, as it can safeguard her future offspring.

Among our own societies, violence toward children is taboo. Not so in baboons: infanticide—the killing of young, usually by males— is rife. In extreme circumstances three-quarters of a troop's offspring can suffer this fate before they reach independence. Males are even known to kidnap infants from their mothers during running battles with other males. If their rival is the hostage's father, the instinct to continue the aggression has to be weighed against the risk to the youngster. With all this heedless macho violence, any offspring that a female produces are in jeopardy. But if she mates with more than one male, the issue of paternity becomes confused, and so, in theory, any males who see themselves potentially as the dad might think twice before harming the infant.

The alpha is usually found in the heart of the group, keeping a close eye on his kingdom, while the others in turn monitor him, glancing at him every few seconds. Periodically, he flashes his eye- lids, or yawns. These apparently benign displays actually convey a threat—the yawn, in particular, shows off his deadly teeth. Yet for all the ferocity required to ascend to alpha status, the boss, once in place, is seldom vindictive or tyrannical. With a powerful alpha and a stable hierarchy, a sense of relative calm settles on the troop. But it can't last. As with so many mammals, the males leave home as they mature, which for baboons means at around eight or nine years old. Older males, too, may switch groups if they think they might get a better billet elsewhere. These itinerants roam the savannah in search of a community to join, and the arrival of one of them at the troop can cause a terrible upheaval. The stranger might try to ingra- tiate himself with the females, smacking his lips together and giving low grunts to signal his benevolent intentions. This does nothing to mollify nursing mothers, who scream their disapproval. Likewise, the resident males, threatened by the newcomer, show signs of agita- tion. The scene is set for noisy conflict, running battles, and injuries, sometimes fatal ones.

A large and powerful immigrant male could even represent a threat to the alpha. The efforts the chief has invested in a succession of battles over weeks or months to reach the pinnacle of baboon society can all be undone in a few short hours. Even if he faces this new threat down, his occupancy of the top spot is likely to be short—just a few months, or a year or two if he's lucky. This is a problem: his legacy is at risk. Late in his tenure, the alpha bears responsibility for the protection of a large cohort of young, his own offspring, who depend on him for their lives. The ascendency of a new alpha can trigger a killing spree. Baboon pregnancy lasts for six months, followed by another year of nursing before the infant starts to take its first tentative steps toward independence. While they are carrying or caring for their young, the females will not be receptive to the new alpha. So for him, the existing young represent a problem—one that can only be solved by killing them to bring their mothers into season. The deposed, former alpha will defend them, if he can, but the butcher's bill in a baboon troop at this time can be devastatingly high.

THE GENTLER SEX?

The social world of female baboons stands in marked contrast to that of males. Even to a casual observer, their interactions lack the supercharged aggression that characterizes the dealings of the males. But a closer inspection shows a rich network of relationships and connections, rivalries and feuds. For males, their birth group is home only until they grow old enough to seek their fortune elsewhere. As males come and go, females remain in the same troop with their family for life.

It's for this reason that females form the core of baboon society. Within the troop, several matrilines coexist. These are extended family groups of related females—grandmothers, mothers, sisters, and their dependent young. These groups within the group are incredibly tight-knit, remaining in close contact day and night for years. The strength of their bonds is affirmed by the exhaustive effort they put into mutual grooming—it isn't unheard of to spend five hours a day burnishing one another's impeccable fur. The kin alliance pays off when squabbles break out between females within the troop. Any female going on the offensive would be wise to ponder her actions first, because picking on one victim means picking a fight with her extended family. As well as bolstering one another in disputes, they even seem to provide comfort when one of their number dies. It's known that bereavement produces severe stress for the family of a deceased relative. Female baboons deal with this in much the same way that we might, by drawing closer to their kin. The matriline group thus acts as a buffer against the turmoil that occasionally afflicts baboon society. The value of this support network shows in the longer lives of females with healthy kin bonds than of those without such healthy bonds as well as in their greater success in raising offspring.

As close as the matrilines are, each member of the group knows her place within the baboon's stratified, rather feudal society. Female rank within the troop isn't earned, or based on ability—it's inherited. Daughters slot straight into the hierarchy one place below their mother and, with maternal support, they can boss every female that they outrank. Just occasionally, one of the lower-ranking females might so far forget herself as to clip a particularly annoying upper-class youngling around the ear, but she'll be careful not to do it when any of the juvenile's family members are nearby. It all looks distinctly unfair when you see a baboon working hard to feed herself only to have an aristocratic female from one of the top matrilines strut over and displace her. This is all the more surprising on occasions when the aristocrat is physically no match for the female she supplants. It doesn't matter if you're a scrawny, shrunken old thing, or a callow youngster: the lower ranks know they must defer to you because this is just how baboon female society works.

There might be half a dozen or more matrilines within the troop, each occupying a position in the hierarchy. Rank inheritance means that a matriline's position in the pecking order can remain unchanged for generations. The females in the most privileged matrilines, unsurprisingly, seem to do best out of the deal: their young thrive and grow faster than the offspring of mothers occupying lower positions. You might think that the lower orders would have a particular interest in change. The problem is, baboon society is very conservative. When conflicts flare among females, the troop comes down firmly on the side of the higher-ranking individual, sitting on the sidelines and chuntering their support for her, sometimes even joining in. This fact of the matter of course makes it very difficult for any would-be Cinderella to escape the fate dealt to her by birth. If fighting your way to the top is unlikely to succeed, then kowtowing and diplomacy can work to some extent. Ambitious females from the lower matrilines try to curry favor by grooming those above them—what else?—and this does seem to have some value in reducing aggression toward the approval-seekers. Oddly, given the importance of family support to female baboons, it is sometimes those who've lost that support who can prosper. Freed from the expectations of joining the family firm and inheriting a (humble) position in society, orphans must rely on their wits and competitive ability. This doesn't always work out, of course, but it's one path that females can follow to escape a predestined low status.

If the females at the bottom of the matriline seem to get such a bum deal, and if there's little they can do to change this, then this elicits the question, Why don't they just leave the troop? After all, this is what males often do if their options are limited. Well, sometimes they do. Large troops can split along family lines when times are hard, with different factions each going their own way. Most often, though, the troop stays together. Part of the answer lies in the trials and tribulations of existence for African wildlife. Baboons are vulnerable to a host of predators. Lions and hyenas are a persistent threat as baboons roam the savannah; crocodiles lie in wait at watering holes; nocturnal leopards are the stuff of nightmares. To escape them, the range of many troops includes a rock face with an inaccessible refuge for sleeping. Even so, leopards present a terrible threat.

I was told once of a foolhardy experiment conducted by a pair of young park rangers in Kruger National Park in South Africa. They each took a leopard pelt from the field station where they were based, and then, for the benefit of a watching audience of people, donned the skins and crawled on all fours toward a troop of baboons. An onlooker that day told me that the baboons' reaction was electrifying: they screamed and sprinted away from the disguised rangers. But having put a little distance between themselves and the pseudo-leopards, the baboons changed strategy. The largest males turned and started to throw rocks at the rangers; then, picking up fallen branches, they led a countercharge. As the baboons closed in, the rangers were forced to abandon their disguises and dive for the safety of their vehicle. They were lucky—in their heightened state, the baboons could have seriously injured or even killed them. If a troop corners a leopard during the daytime, their ferocity is such that they occasionally mob it to death. All of this means that there's safety in numbers, and the presence of the larger males in the troop especially provides a level of protection that would be lost if a matriline of females were to leave.

Another reason to stay is the risk of infanticide. A group of unattended females would most likely attract the attentions of a male, which would immediately imperil any nursing youngsters. While this is also a risk within the troop, females are able to count on "friendships" with male troop members to balance it. As we've already seen, during pregnancy, and while they're weaning their young, females aren't sexually receptive. Regardless, they seek out a male and invest plenty of time in him, following him around and grooming him whenever they can. In some cases, it may be that the male is the father of the offspring they're carrying; in others, he might not be, but is tantalized by the prospect of being her partner the next time she's ready to mate. Sometimes the male is neither the father of the baby, nor about to be the father of the next one. It's hard to see what this male gets out of the relationship, other than a few less ticks from the grooming he receives and a good reputation (or one as a sucker) among the troop's females.

From the female's perspective, having a consort—a knight in hairy armor—as an ally is priceless when a change occurs at the top of the male hierarchy. If the male hears the distress cry of a female with whom he has formed a bond, he will rush to her aid, even though doing so might put him in the sights of a powerful, dominant male. Not all consorts are equal, though, and the value of a high-ranking male puts a premium on his services. If there are lots of females vying for his protection, jealousy can rear its ugly head. If a flagrant jezebel comes flirting around her male friend, a female who's already with child risks a dilution of his efforts in the future. For this reason, females with a high-ranking consort are hostile to the approach of new females, driving them off or bullying them to the point that the stress makes it difficult for them to conceive.

Even though, as in most primates, baboon mothers carry the responsibilities of childcare, the fathers do play an important, if discreet, role in their offspring's life. This might seem strange in light of the relentless, hectic aggression of male society, but while male baboons are certainly not in line for "Dad of the Year," there are measurable benefits to the young in having him around. If a playground scrap develops between two juveniles, adult males often step in on the side of the younger one. They're especially forthright if it's their kid in trouble. The males will also use their power to boost their offspring's chances of getting a fair share of food resources. In *Baboon Metaphysics*, an entrancing account of a lifetime of baboon research, Dorothy Cheney and Robert Seyfarth painted the charming picture of an old male, long since having fallen from the grace of being alpha, followed by a trail of youngsters whose destiny he had helped to shape. All in all, while it may seem that baboons at the bottom of the hierarchy might be better served by jumping ship, the ties that bind in baboon society are strong.

For baboons to prosper in their social world, they must have a sophisticated understanding of the relationships within the troop. Like us, they recognize individuals by sight and by sound. They can distinguish their relatives from unrelated individuals, too. This isn't so surprising in the context of their maternal relatives, because they are typically reared together. Yet they can also identify troop members who share the same father—in other words, their paternal half sisters or brothers. Through this recognition, they can build their networks within the troop according to relatedness. However, navigating their society requires much more than this. It's essential to know who's related to whom, or at least *cozy* with whom. This might partly explain why baboons, along with so many other group-living primates, are so fascinated when there's a birth in the group. It's hard to overstate this natal attraction. The mother of a new baby can receive over a hundred visits a day, particularly from other females who have their own newborns. Most of the time, the visitors want to handle the baby. And a high-ranking visitor will insist regardless of the mother's misgivings.

From the first days of their lives, individual baboons are part of the social fabric of the troop. When this fabric threatens to tear as the result of a falling-out, reconciliation—in females at least—tends to follow fairly swiftly. It takes the form of a grunted reassurance from the dominant party in the dispute. This response assuages a nervous subordinate's fears and tells her that all's well. Sometimes the dominant's relatives do the reconciling on her behalf; it serves no one's interests to have a long-running feud. Experiments where recordings of particular individuals' calls were played in the vicinity of others have provided some valuable insights into how well baboons are tuned into troop relationships. For instance, when baboons hear a juvenile's frightened scream, they look at its mother; when a subordinate hears the relative of a dominant quarreling, she looks at that dominant; and when two baboons hear a playback of their respective relatives quarreling, they look at one another. This kind of response doesn't happen when baboons hear the calls of unrelated animals. It demonstrates just how well they understand the myriad connections in their network.

This combination of intelligence and societal know-how in baboons has been put to use by our own species in remarkable ways. There's the story of Jack the baboon, who worked as a signal operator on the Cape Railway in South Africa for nine years toward the end of the nineteenth century, assisting a man who'd lost his legs in an accident. Jack received a small salary along with a treat of beer on the weekends from the railway and was famed for never having made a mistake in all that time. More recently, a German naturalist, Walter Hoesch, described a baboon who was being used as a goat-herd in Namibia. Ahla, as the baboon was known, was excellent at the job—certainly better than people, according to the farmer. She would keep the flock together, unerringly noticing if one of almost a hundred goats went missing. She'd round up the strays when needed and would call out if a predator threatened. At the end of each day, she'd muster the goats with a "ho-ho-ho" call and escort them back to the protection of the kraal, or fenced enclosure, riding on the back of one of the rearmost goats in the same way that an infant baboon often rides on its mother's back like a miniature jockey. As the goats packed into the kraal, kids and mothers would inevitably get separated in the confusion. Ahla would remedy this by carrying each kid to its mother under her arm, flawlessly identifying which youngster belonged to which mother. Although this is the most well-documented case, the use of baboons down on the African farm has a long history. The Scottish explorer Sir James Alexander reported that the Nama people used baboons as herders in the 1830s. Ahla's ability to identify each individual goat and to recognize the relationships between kids and mothers was an extension of the skills that all baboons must master to thrive in their society.

FINDING OURSELVES

Chimpanzees are our closest living relatives. The overlap between our genetic material is startling, with some estimations suggesting almost 99 percent similarity. Among all animals, then, do they offer us the best chance of understanding the evolutionary roots of our society? Are we like chimps? Are they like us? To examine this in more detail, we need to understand our ape cousins and how they live. While many excellent studies consider captive groups of chimpanzees, the best data comes from observing them in the wild. The problem is that studying chimps in their natural habitat is a major challenge.

To start, the human observers must very gradually acclimatize the chimps to their presence in a process known as *habituation*. Gradually is an understatement—this can take years. Even after this, the observers must be unobtrusive, so that the chimps behave as normally as possible. Another crucial ingredient is perseverance: a good data set needs thousands of hours of observation, often across several years. Jane Goodall's pioneering studies of wild chimpanzees nearly sixty years ago at Gombe Stream on the shores of Lake Tanganyika have inspired other researchers across the continent, from Guinea and the Côte d'Ivoire in the west to Uganda and Tanzania in the east. Their work has transformed how we look at chimpanzees and in turn has shaped our understanding of our own evolutionary story.

What we've learned about chimp society has at times been unsettling, both to our rose-tinted view of chimpanzees and because of our close genetic relationship to them. This was never more so than in Goodall's account of a staggeringly violent turf war between two chimpanzee communities in Tanzania. The conflict took place over four years in the mid- to late 1970s, and Goodall's vivid account of kidnappings, beatings, and killings were so shocking that they were initially viewed with skepticism. Her eyewitness report described males who had once shared close relationships tearing into one another, pummeling fallen rivals to death with rocks. Unforgettably, she described three males capturing a female who was fleeing with her infant. They administered a terrible beating to her before mercilessly smashing the infant against the ground and flinging its limp body into the undergrowth. Stories of killings in other communities, including infanticides and the cannibalism of young, have since emerged to corroborate the earlier reports from Tanzania: there's a dark underbelly to chimpanzee life.

The history of humanity is littered with stories of conflict and bloodshed between rival groups. So it is with chimpanzees. Territory is crucial to chimps for the resources it contains, including both the food that they need to survive and, for males, the breeding rights within the community. Outsiders are a threat. To safeguard their claim, males form into alliances and patrol the margins of their land. If intruders, or a border patrol from an adjoining community, are sighted, then there may be trouble. There's a show of strength—noisy posturing, charges and countercharges, the flinging of projectiles. Amid the chaos and raw aggression, the confrontation can turn lethal. Across Africa, researchers have accumulated detailed, long-term profiles on eighteen chimpanzee communities over the decades since Goodall's time. The most recent estimate for deaths by violence among these communities stood at 152. Both victims and perpetrators were most likely to be male, and the greatest cause was fights between different communities. On this basis, it seems that the warlike tendencies of mankind have parallels with our closest living relatives. It's a theory that has been popular in recent times. Perhaps, suggest its advocates, the instinct to kill is embedded in our shared genetic heritage. The idea that violence is written into our DNA has a perverse appeal, because it suggests that the litany of horrific wars that stain our history were predestined, inevitable. It simultaneously offers an explanation and at least partially absolves us of responsibility.

The idea that conflict is in our genes, that we're hardwired for fighting, is problematic, though, because it's a very simplistic view of humans and chimps. We may very well have some violent tendencies, yet these are not the only ingredients of our character, and they're balanced by other, more sociable inclinations. Some people kill for the contents of a wallet, while others freely give blood without reward. How can you summarize a species with such diverse inclinations? The same applies to chimps. Although murderous behavior by chimps is certainly not unheard of, the kind of protracted, community-destroying turf war described earlier is extremely rare. To judge chimpanzees solely on the basis of these events would be to do them a great disservice, just as it would to view the essence of our own nature exclusively through the prism of apocalyptic newspaper headlines. Like chimps, we demonstrably have the capacity for violence; although we shouldn't gloss over it, we have to view it in the broader context of cooperation and coexistence that characterizes our day-to-day lives within our societies.

According to Frans de Waal, a biologist and an eminent scholar of primate behavior, the most striking aspect of chimp behavior is not the aggression in their society, but how they reconcile their differences, and the value they place on maintaining relationships. Chimps literally kiss and make up. They embrace one another. They groom and caress even their opponents in the aftermath of conflict. They show the hallmarks of empathy, exemplified by how they respond to distress in others, offering consolation and a hug to anxious or sorrowful members of their community. The pioneering Russian psychologist Nadia Kohts described her relationship with Joni, a male chimp that she raised at her home in Moscow. Joni would occasionally escape to the roof of the building, where he couldn't be followed. Efforts to get him to come back down followed the usual tactics you might try with a pet cat or dog—Kohts offered his favorite foods as a bribe. When this proved—ironically—fruitless, Kohts tried a different tack, feigning to weep. At this, Joni hurried down from the roof to console her. The more distressed she appeared to be, the stronger the effect on Joni, who cupped her chin in his hand, stroked her face with his finger, kissed her, and made sorrowful noises of his own. Joni's behavior was typical of chimps: they show an attentiveness to the needs of others that is matched by a tendency to forge and nurture relationships, traits that are characterized more than anything else by cooperation and harmony.

Chimp society mirrors our own in many ways. They live in large groups, known as communities. These can include well over a hundred individuals, who will often travel and forage within the geographical boundaries of their community, either alone or in small collectives, known as parties. Individuals from the community frequently meet and either coalesce into parties or go their own way for some time. Chimp society is dominated by the larger, stronger males, who outrank females and who compete among themselves for the privileges of rank—specifically, for extra helpings of both food and sex. Females also have a dominance hierarchy, with position being determined more by age than by the direct contests of aggression that characterize the males' struggle for power. Females usually disperse from the communities in which they were born as they become sexually mature, which means a perilous journey into a new group and the very real risk of attack by the chimps she finds there. It will be a long process before she's fully accepted, and even then, she'll have to adopt a lowly position.

Getting ahead in chimp society requires a mix of diplomacy, strategy, and social maneuvering, which in turn can be facilitated by what, in human society, we might call networking. Grooming is a major part of the lives of many social animals, and chimps are no exception. Among primates, the more sociable a species—which is to say the larger its group size—the more they groom. The time investment can be staggering—some chimps spend 20 percent of their day grooming and being groomed. Not only do they end up magnificently manicured, but they also strengthen their affiliations with their grooming partners. At a deeper level, grooming between cooperative partners stimulates the production of oxytocin, the so-called love hormone, which promotes social behavior. And as a political strategy among chimps, grooming takes some beating. It's become a key part of the chimp diplomatic toolkit, which might explain why males groom one another more than females. Though it's time-consuming, it doesn't risk the injuries that a more aggressive strategy might incur.

Whom they groom is important, as is who's watching. A subordinate may try to curry favor by grooming one of his superiors, but equally, if there's a more senior chimp in the audience, he may well trade up and start grooming that one instead. If the higher-ranking male rejects these overtures, as sometimes happens, this sucking up falls flat. Indeed, grooming can get competitive. When males knock one another out of the way for the privilege of grooming a particularly favored male, we might even recognize something akin to jealousy. Chimps develop relationships through a savvy mix of grooming, cooperation, and hanging out with high-ranking peers. As well as valuing friendship in its own right, they're wily strategists who scheme and finesse their way to status.

MEAT MARKETS

In order to defend their territory, resident male chimps work as a cohesive unit. However, this isn't the only area of chimp life that rewards cooperation. Hunting, too, demands a sophisticated and coordinated approach. Until relatively recently, it was assumed that chimps, like their ape relatives, gorillas and orangutans, were vegetarians. Though most of the wild chimp diet is made up of fruit and leaves, meat provides a tasty and highly nutritious addition. When Jane Goodall first reported her observations of chimps actively hunting animals and consuming meat, there were many who reacted with disbelief—and thought this was a one-off, just a community of chimps gone rogue. In recent decades we've realized that hunting is widespread among chimps. Again, we see in this our affinity to our closest relatives, and as was the case with our own ancestors, hunting in chimp society is largely a collaborative affair. Although chimps can and do hunt on their own, their success increases dramatically when they do it as a group. Most out-and-out carnivores—lions, for example—would be lucky to end one in every two or three attacks with a kill; yet a group hunt by chimpanzees nearly always ends in success.

It's sophisticated teamwork that gives chimps the edge when they hunt their favored prey through the forest canopy. They're generally looking for smaller primates, such as Colobus monkeys. Each individual on the hunt has a specific role—some chase while others take positions to block the prey's escape and funnel them into a trap; still others wait concealed, ready to ambush their unfortunate victim. Hunting parties like this are largely a male affair, and the spoils of the hunt are largely shared among the participants. Despite their agility and strength, adult females are in charge of childcare. Carrying an infant makes them far less mobile than an unencumbered male, which pretty much rules out the possibility for them of charging around in the treetops after highly mobile monkey prey. In some chimp communities, at least, females nevertheless do hunt. On the savannahs of Fongoli in Senegal, females actually fashion spears from tree branches, carefully shaping the weapon by stripping away side branches until they're left with a formidable stabbing tool. Aside from our own species, this is the only known example of predators using tools to hunt large prey. Armed with their spears, the chimps will seek out a nocturnal primate: the bush baby. These inoffensive little animals—distant relatives of the chimps—sleep through the day in the safety of tree hollows. But if a chimp finds one resting, it skewers it and retrieves its prey from its hidey-hole.

So far, we've seen that chimpanzees cooperate in the defense of their home turf and that they combine to hunt agile and evasive prey. The meat that results from these hunts is nutritious and incredibly valuable. So much so, in fact, that it becomes a currency in chimp society. Although chimps don't usually share smaller fruits or other parts of their diet, they do share meat—but not randomly. Dominant males use it as a reward for their male supporters, as well as giving it to particularly favored females. This meaty patronage shows its worth in family life—more meat in the diet translates to more surviving offspring. In this behavior, we get a sense of the complex society of chimps. While the dominant male may very well be tough, and he may have had to prove himself physically, it's the size of his support, rather than that of his biceps, that will allow him to maintain his position. Rewarding loyalty with meat is part of his strategy.

Mammals aren't the only meat, of course. Chimps across Africa collect insects such as termites and ants from their nests. Small though they are, the insects are packed with nutritious fats and proteins to bolster the chimpanzees' diet. Ant and termite nests are well defended, however. Their durable structures house batteries of soldiers who will rush to defend their colony and can inflict painful bites. Capturing the insects requires a deft touch. First, you need the right tools for the job: one to puncture the nest and another to extract the termites. Although many chimps go fishing for termites, some have elevated it to an art form, fabricating a specialized tool. They make it from stems, and they're choosy about which plant they will use. It turns out that arrowroot fits the bill nicely. The chimp strips the spindly stem down and then nibbles it at one end, separating the fibers to make a kind of brush. This can then be inserted into the hole they've made in the termite nest, and the agitated soldiers unwittingly do the rest, biting down and attaching themselves to the invading brush. It's then a simple job for the fisherchimp to extract and eat them.

Once a group member masters a new skill, its admiring neighbors can then copy (or "ape") its behavior. When chimps move between groups, they sometimes bring valuable information with them. Ant fishing was unheard of in Gombe until an immigrating female brought the tradition from her birth group. Her novel techniques spread quickly, especially among the young—perhaps the more senior chimps thought that dipping for creepy-crawlies was beneath them. In a relatively short space of time, ant fishing became a firmly established pattern of behavior within the Gombe community.

MACHIAVELLIAN MANEUVERINGS

Chimpanzees are a living example that shows that when it comes to leadership, size and power don't always matter. One smaller male rose to become the alpha male in his community by dint of the sheer hard work that he put into building relationships through grooming. When we talk about "alpha males" in human society, it implies a certain ruthlessness and a tendency to dominate. In animals such as chimps, for whom the term was originally coined, the alpha male may be the leader, but he can only rule with the support of his community. He must be a relationship-builder rather than a despot.

As in human society, political games aren't just limited to the chimpanzees' chief. Even as the alpha chimp seeks to expand his power base, his rivals may be attempting to secure their own alliances, especially if they hope to mount a challenge to displace the leader. In the shifting diplomacy of chimp society, it pays the dominant male to keep a close watch on the various intrigues and affiliations as they develop, especially as allegiances can be tenuous and short-lived. If a male chimp decides his interests will be better served by switching to another clique, he'll do so. Low-ranking males, in particular, have a stake in trying to change the status quo, since they receive low rations of all the things that make chimp life worthwhile. So these floating voters may often swap sides. Another problem the dominant male faces is that a chimp community can be dispersed over several square miles, so his rivals may not even be within hearing range, let alone in sight. When he does encounter his opponents, it's useful for him to remind them who's in charge. He does this by means of an awe-inspiring display of screaming and charging about, bristling with intent, though not necessarily a violent one. His supporters may be reassured, while his competitors either lie low or display their subservience—sometimes through a vocalization that is known as the *pant-grunt*, normally given by subordinates to their superiors.

In recent times, Internet memes of politicians before and after they reached the pinnacle of their careers have gained popularity. Among them we can see the physical toll the presidency took on a once luminously fresh-faced, dark-haired Barack Obama; similar before-and-after comparisons have been made with Tony Blair and Angela Merkel. In chimp society, being the alpha carries plenty of prestige, and more tangibly, it means plenty of food and sex. But nothing lasts forever. Even while he's in charge, the responsibility is taking its toll. The stress that leaders experience is measurable in elevated concentrations of the hormone cortisol. Although this and related hormones have the benefit of keeping the leader on his toes and ready for action, raised cortisol levels over time have a whole host of damaging effects, including weakening the immune system, disrupting sleep, and triggering muscle wastage. Signs of sickness or injury in the alpha or in his immediate supporters will be monitored by those in the lower ranks, and any perceived weakness may lead to a challenge. In natural chimp communities there will always be pretenders to the throne, and as females, not males, are the dispersing sex, there are always young bucks coming through the ranks with time on their side and an eye on the prize. As a rival faction grows in strength and numbers, communal tensions build. Faced with a challenge, it is up to the alpha and his allies to put it down—his reign lasts only so long as he is able to see off the opposition. The alpha might be in charge for over a decade (more usually between three and five years), but eventually the day will come when he is overthrown. This is a traumatic time for his community. Occasionally, the intensity of the rivalries can lead to death.

A detailed, long-term study of chimp communities in the Fongoli woodlands of Senegal shed fascinating and troubling light on the fate of one particular alpha after a coup. Central to the story is a male the researchers named Foudouko, who rose to become the alpha male in his late teens. Foudouko held on to his position for around two and a half years, supported by his second-in-command, known as MM. Though it's not certain what led to Foudouko's overthrow, MM had suffered a serious injury in the lead-up, which suggests that, stripped of his closest ally, Foudouko became vulnerable. In any event, he was deposed. Finding himself at the social periphery, he withdrew almost completely for five years. This is exceptionally unusual for chimps, especially males, though signs gradually started to appear that he was being rehabilitated into the group following his long exile. His bond with MM remained strong, and the new alpha, MM's brother, appeared to be tolerant toward him. Not so the other males—maybe they had a score to settle from years before, or were disturbed by the prospect of Foudouko displacing them in the hierarchy. One night, shortly after his reappearance on the social scene, an intense fight broke out.

The following morning, Foudouko's body was found. The violence that had been meted out was seemingly vengeful as well as fatal, since his former neighbors kept attacking his corpse well after his death. Some were even seen eating parts of his flesh. Notably, MM and the new alpha didn't join in; MM, in particular, seemed to be trying to protect him and at one stage to revive his dead friend.

The violence triggered agitation that rippled through the community the next day. Some members showed signs of nervousness, while others still displayed the pent-up rage that had led to the killing. The hyperaggressive behavior that the Fongoli males exhibited might have been due, as in the chimp war described earlier, to a relative shortage of females. When Foukoudo began his reintegration to the community, tensions between males over mating opportunities may already have been high, and the addition of another male only made things worse. A balanced sex ratio among chimps is critical to keeping the peace. It's for this reason that the actions of poachers can have far-reaching effects. Often, the poachers target females, especially those with infants who will command a high price in the illicit pet trade. The net effect of this is an imbalance between males and females that can destabilize an entire community.

So far, this has been a male-centered account of chimp society, reflecting the fact that theirs is a male-dominated world. The females are outsized and outranked by the males, and their contribution to the intrigues of male chimp politics is less immediately obvious. Yet their support of particular males can be important in deciding who gets, or keeps, the top job. The sexes each have their own pecking order. While there's drama in the competition between males for the privileges of rank, the female hierarchy is established in calmer fashion, usually as the result of an orderly queue, decided by age and experience. Females are also the less sociable sex, and it's often said that their relationships aren't as strong as those between their gang-forming male counterparts.

The importance of kinship to the favoritism often shown in animal groups is well known. The theory has it that since female chimps aren't so closely related to each other as the males, they're also less invested in one another. Yet female chimps don't always conform to expectations. Living together for years, they form lasting, powerful relationships. This tendency sometimes shows up as solidarity in the face of a threat. If a male decides to try his luck and impose himself on a female, especially on a high-ranking one, he risks reprisals from her allies.

Before we rush to the conclusion that the females are the more civilized sex in chimp society, it's necessary to paint the full picture. When a young female tries to join a new community, she faces a period of adjustment. The males are no doubt delighted to welcome such a newcomer, the females not so much. The latter have been known to band together to give an immigrant female a beating and drive her off. This intolerance from the resident females is a real challenge for a new arrival, and sometimes her only option is to seek the protection of one of the males—so much for sisterhood. Even if she does get to stay, the young immigrant has to be satisfied, at least at first, with one of the lower rungs of the social ladder. After that, well, it's not all plain sailing. The matriarchy of established females can be cliquey. They monopolize the best feeding spots, relegating the lower-ranked females to the margins. It doesn't stop there, either. If they're sufficiently disturbed by a new arrival's continued presence, this can lead to further attacks on her and any offspring she has, up to and including killing the infant. For these reasons, low-ranking females must be cautious. They keep away from other females, especially when the time comes to give birth and they are especially vulnerable.

Unsurprisingly, the competition between male chimps and between female chimps is ultimately about scarcity. Food is one part of this problem, but it isn't the only resource, a fact that leads us neatly, if nervously, into the murky world of chimp sex. From a female chimp's perspective, the arrival of a new girl in the community means both another mouth to feed and more competition for male attention. Conversely, access to females is a major reason for the feistiness between males. Charles Darwin suggested long ago that the sexes should have different strategies when it comes to mating. Females usually invest more in reproduction, so they should be more choosy, while males, as relatively low investors, should compete with one another to mate with as many females as possible. How does this play out in chimps? Well, chimp pregnancies are nearly as long as human ones, at eight months, and afterward, the female shoulders all the childcare responsibility, so females make by far the greater investment.

Interestingly, chimp females are picky about mates, and not necessarily in a way you'd expect. In fact, they seem to play a very smart game. Their menstrual cycle lasts for slightly over a month, and for about a third of that time, they're sexually receptive. In common with lots of other primates (but happily, not us) who live in groups where more than one male is present, they bear a very visible sign of this availability. Their whole genital area swells with water to become massively—shockingly, even—distended. There's nothing coy about a chimp in this state: she'll have sex with most, if not all, of the males in her group. There is lots and lots of sex. When the chimp female reaches the critical time when she's most likely to conceive, she switches her strategy and focuses her wiles on the alpha male. But mating with lots of males while she's unlikely to conceive is a good way of currying favor with them. It may also put enough doubt in the males' minds as to the paternity of future offspring that the chances of them attacking or even killing the infant are reduced.

When seeking mates, female chimps choose dominant males. No great surprise there: it's a common strategy and a valuable one for females of many species, as it usually means good genes and sexy sons. But what about the males: What do they find sexy? You might be expecting to hear that they're entranced by young, shapely females with the chimp equivalent of a well-turned ankle. Not a bit of it. They won't pass up an opportunity to mate if one arises, but, given a choice, what they like best is an older, heavier lass who's had lots of kids before. Again, it makes sense, even if it's slightly surprising, because such a female is likely to be a better forager and will also be a more experienced mother. She's likely to stand higher in the dominance hierarchy, too.

Although both sexes have their strategies, that doesn't mean they necessarily have the freedom to pursue them. This is a difficult topic, not least because we're talking about a species so similar to us. We can't duck the issues, though: chimps and many other group-living animals experience sexual coercion. In chimps, lower-ranking males seek to cancel out female preference for dominant males by aggressively pursuing them for sex. As females don't necessarily put up a strenuous fight, it can seem as though they accept this—but because of the imbalance in size and strength between the sexes, unless she can solicit help, there's not much she can do to prevent it. On top of this, not only the guilty male but the female, too, can end up being punished by higher-ranking males for their promiscuity. For this reason, females seek the company of the dominant males as a measure of protection when they're at peak fertility. It's a good strategy for the fertile female, perhaps, but not for the other females, who may seek to throw a wrench in her works. Females have been known to separate mating pairs in order to have sex with the male themselves and thus deprive another female of her conjugals. Maybe that's why some females are sneaky.

In the strange world of chimp sex, females often produce a particular sound, a kind of screeching noise known as a copulation call, which serves to advertise their availability and to encourage males to compete for her. If there's a dominant female in the neighborhood, however, lower-ranking females keep their liaisons with particularly desirable males secret, having sex in discreet silence.

WOULD YOU WANT TO BE A CHIMP FOR A DAY?

I don't know if you've ever pondered which animal you'd like to be if you weren't human—maybe it's a biologist thing. Of all the people I've asked, no one has ever said chimp. I've had wannabe eagles, lions, tigers, sharks, dolphins, whales, and even one vote for a sloth. Not once has anyone said a chimp. Maybe that reflects the taint that attaches itself to chimpanzees now, our view of the sometimes brutish world that they inhabit. If that's your view, well, it's understandable, in the light of some of what's gone before in this chapter. But it's worth examining chimpanzees in the round, and considering some of the better aspects of their nature.

Living alongside one another for extended periods, exhibiting the intelligence to recognize other individuals, form opinions of them, and potentially understand them to a degree far greater than just about any other animal, chimpanzees are the consummate social beings. Experiments using bartering systems, where primates can choose to exchange tokens for food, can tell us things about their sociality. This has been tested in chimps in a similar way to capuchins, as described earlier, where the capuchins traded pebbles for food.

In an experiment with chimps, one token bought them a piece of carrot, and the other a grape. Like the capuchins, chimps favor the grapes. But this time, a high-ranking female chimp was trained to buy only carrots and not grapes, even though, given a free choice, she would have preferred the grapes. Surprisingly, though, when lower-ranking chimps saw her buying carrots, they fell into step with her, copying her preferences over their own.

You might conclude from this that here's proof that chimps are chumps. Let's pause and think about that for a moment. Have you ever been in a discussion where you've ended up agreeing with someone who holds a different opinion from yours? Have you ever been influenced by someone else's behavior? If you haven't, then well done—you're just about unique. For the rest of us, conformity—much as we might rail against it—is part of the glue that holds societies together. Chimpanzees may copy the actions of a dominant figure because they expect experienced individuals to make good choices. Or because high-ranking chimps are influencers—in other words, because the lower-ranked chimps want to be like them, and to be liked by them. Either way, the tendency to conform acts as a unifier in their society, just as it usually does with us.

And what about conflict within their communities? True, it can spiral from arguments to macho bravado to violence, even murder. Far more often, though, chimps weigh their temporary antagonism toward an individual against the possibility of long-term future cooperation. They don't like to burn their bridges rashly. Reconciliation is common following disputes, and amicable relationships are central to their social life.

Competition remains a fact of life for all animals. In a world where the most coveted resources are scarce, there's seldom enough to go around. Each animal—each chimp, in this case—has to look out for itself. In spite of this, they show an amazing ability to cooperate. The thing about cooperation is that there's always the risk that some individuals will cheat. Doing so is a pretty successful—if immoral—short-term strategy; the cheats get something for nothing. The better overall strategy, though, is teamwork. And for cooperation to work, you need to have some punitive measures for those who don't play fair.

The disapproval of peers and the risk of reprisals that may follow cheating provide means to manage it. Comparative psychologist Malini Suchak and her colleagues studied this issue at the Yerkes National Primate Research Center in Georgia. Groups of captive chimps in large enclosures had to work out how to solve a simple task. The trick was for pairs or trios to work together, coordinating and synchronizing their efforts. Group members were free to decide whether to engage with the task and for how long, and they could choose whom to work with. If they were successful, they triggered the release of small pieces of fruit. When this happened, the food could be taken by the ones who had solved the task (i.e., who cooperated) or by a sneaky onlooker (i.e., a cheater). As soon as one batch of food was released, it was replaced by another, so the chimps could repeat the task right away for another treat. They could do this as many times as they liked over the course of an hour. The procedure was carried out episodically for one hour at a time a couple of times per week over a period of several months, which allowed the experimenters to see how the chimps adjusted their tactics over time. The question was, would the chimps follow a cooperative strategy, or a cheating one?

The results vividly demonstrate how the interplay between chimps defines their strategies. Having figured out the task, the chimps got to work, collaborating successfully to solve it and so release food rewards. Initially, cooperation was the winner—the apes that put in the effort got the rewards. However, as they became wise to how the system operated, some took the opportunity to freeload, snatching the food as it appeared without putting in the effort. The cheating strategy became more and more prevalent, and cooperation started to decline. But, as in the wild, cheating chimps got punished. How the cooperators punished the cheats varied, ranging from sulking and refusing to activate the equipment to angry displays of justifiable annoyance at this antisocial behavior. At some time or other, all the chimps in the study had a go at freeloading, yet only one turned out to be a serial offender—an elderly, blind female who became something of an outcast as a result of her cheating.

Cheating's a bad strategy when there's a price to pay for it. In the experiment, just when it looked like cheating might prevail, the cooperative chimps got to grips with the freeloading and started to police it effectively. From there on, cooperation went from strength to strength. When Suchak ran the experiment again with a different group, the outcome was pretty much identical. It turns out that chimps are natural team players.

It was once thought that empathy and compassion were uniquely human characteristics. Foudouko's violent death and his community's ruthless response give some credence to this—there's not much evidence of finer feelings there. Yet that kind of response seems to be no more typical among chimps than, say, outbreaks of mob violence are in our own species. More typically, chimps mourn the death of a community member. Geza Teleki, a researcher working with Jane Goodall at the Gombe Stream Research Station in Tanzania, saw a chimp break its neck and die from a fall. Like Foudouko's death, the event triggered manic and aggressive activity among the fallen chimp's peers. But this soon settled into apparent concern for one another and for the dead chimp. They embraced one another, perhaps for reassurance, and, approaching the body, touched it gently. They then stood gazing at it while either whimpering softly or keeping an un-chimp-like silence. Other descriptions corroborate the uncharacteristic solemnity that affects chimps following a death. They touch the body as if in contemplation, remaining there for some time and suspending normal chimp activities. As with us, they seem to grieve most for the loss of a close friend, returning to the body time

and again, and sometimes defending it from the attentions of boisterous youngsters. Moreover, they show concern for the sick and injured. Jane Goodall's accounts of the chimps of Gombe, which gave
us so much early insight into all areas of wild chimpanzee behavior,
include many instances of chimps supporting sick and injured colleagues. A young male cared for an elderly friend in the last weeks
of his life, diligently defending him against the curiosity of much
higher-ranking chimps. An adult male's screams of pain brought his
aged mother to rush from a third of a mile away to console him.

In his book *Good Natured*, de Waal describes an old male whose
alpha status was slipping away in the face of a challenge from a
younger and stronger rival. There was little he could do, of course,
since sooner or later one generation must give way to the next, but
his anguish was clear. He would charge around, noisily beseeching
others to help. It fell to the females to approach, wrap an arm around
him, and soothe him. On the whole, chimps are a captivating and
enigmatic mix of brutality and compassion, of altruism and selfish-
ness. We see in them a complex blend of opposing characteristics, a
blend we might well recognize in ourselves.

BIG BRAIN, BIG HEART

The vast Congo River winds through the heart of Africa, carrying
more water than any other in the continent and draining an area
larger than India. The scale of the river is breathtaking—the far bank
may be as much as 8 miles away. The Congo rises in the hills around
Lake Tanganyika and northern Zambia and flows north before coming around in a great arc, crossing the equator twice as it does so,
then heads south and west to the Atlantic. It is flanked by the greatest rainforest in Africa, a rich, fertile, steamy expanse of green, the
cradle of tens of thousands of species. The modern course of the river
presents a formidable barrier, a natural moat, between the animals to
the north and those to the south.

At some time in the long-forgotten past, between one and two million years ago, a band of chimp-like apes took advantage of extreme

conditions—perhaps a major drought—to cross the river from the north. They made their home in the southern lands beyond the river, free from competition from their own kind. There, these bonobos, as we call them, have stayed ever since, separated and protected from the chimpanzees to the north. Both species have a strong dislike of water, so the Congo River is a particularly effective border between them.

Genetic analysis tells us when the bonobos separated from chimpanzees and also suggests that there was perhaps only one major river-crossing event. If that's the case, it was an extraordinary, pioneering event. Since that first crossing, it seems that there has been the occasional refugee from one side or the other, enough to allow just a soupçon of interbreeding, but not enough for bonobos to establish north of the river, or for chimps to establish to its south. Even now, there is little difference between the DNA of the two species—their genomes differ from one another by only about 0.4 percent. So similar are they, in fact, that primatologists didn't distinguish between them as separate species until 1933. Bonobos remain the most recent addition to the handful of ape species recognized in the field.

So how would you tell a bonobo from a chimp? Bonobos are slightly smaller, for one thing, with longer limbs and a smaller head—which doesn't quite warrant their former name, the pygmy chimpanzee. The fur on their heads is longer than in chimps, which contributes to a rather marvelous hairstyle that brings to mind a Victorian gentleman, sometimes with a center parting, other times chaotically boffin-like. Their appearance aside, the two species have much in common. They eat similar foods; live in large, mixed-sex societies; and are both supersmart. Female chimps and bonobos both move on to new pastures when they mature, and they're smaller than their male counterparts in both species. You could be forgiven for thinking that there's really nothing to see here: two similar species, doing similar things. And yet you'd be wrong. The differences between them are striking, and with their close evolutionary relationship to us, there's a great deal we can learn about our own origins by studying them. Some of the very best evidence we have comes from the bonobo, an overlooked animal that is just as closely related to us as the more prominent and celebrated chimpanzee.

Still, it took a long time for science to get up to speed on this. Compared to the attention that's been lavished on the study of chimpanzees, bonobos—until relatively recently—have been rather neglected. This might have to do with lazy latter-day assumptions that bonobos were just chimps writ small, or that they're far less widespread and less common than chimps—and it is true that there are only about twenty thousand of them in the wild and very few in captivity. Their home in the Democratic Republic of the Congo has a tragic history of conflict and political instability, making study of them a major challenge. Yet over the past three or four decades, our understanding of our forgotten cousins has improved dramatically, and we now appreciate them for the illuminating creatures that they are. A word of caution, though—biology isn't always family-friendly, for a range of reasons, and that's never more so than in the case of these apes. It's important to study and to mention these aspects of their behavior, though, because it shapes the dynamics of their society in a fundamental and unique way.

Bonobos are very sexually active. Very. They have sex to say hello, to release tension when they're excited, and to make up when they fall out. Uniquely, they French-kiss—with tongues—and they sometimes have sex in the missionary position. They don't confine themselves to male-female couplings, either. Pretty much anything goes. Female bonobos have startlingly large clitorises and pair up to rub their genitals together several times a day, shrieking with the thrill of it. Males rub their penises together, sometimes inventively dangling from branches to fence with them. They have oral sex and finger fun. It's thought that females might even make dildos. It's perhaps not surprising that this element of bonobo behavior has attracted the widest attention, given our human obsession with salaciousness. The danger is that it obscures the wider picture and leads us to think of bonobos as lascivious, sexually obsessed miscreants. To some extent, they *are*, but they are a lot of other things, too.

Bonobos are far less aggressive than chimps, and part of this stems from their use of sex to discharge tension. While chimps may use aggression to get sex, bonobos use sex to diffuse aggression. This all sounds very creative, rather like the counterculture slogan of the 1960s: make love, not war. Yet the prominence of sex in bonobo society is an indication in itself that there's some aggression to control or allay; they just manage it differently from chimps. One of the suggested underlying causes of male aggression in chimps is sex, or the lack of it. Chimp females are sexually receptive only when they're in estrus—the very visible signs of it are shown in their swollen pink bums. Estrus only begins with sexual maturity, of course, but they also don't come into estrus for years once they give birth. It's estimated that, across her whole lifetime, a typical female chimp is only sexually receptive for about 5 percent of the time. In chimp communities, this means that there are very few receptive females at any given time and that, hence, there are lots of frustrated, angry males.

Female bonobos are in estrus about five times more than their chimp sisters. They even show something known as pseudo-estrus: they sometimes appear receptive—in the pink, so to speak—even when they're not ovulating. From a male perspective, this means far more sexy time, and far less strife. From the female perspective, it boils down to far less pestering. In chimps, females can do little to ward off concerted male attention and may even suffer severe attacks. This just isn't part of a female bonobo's experience. Whether or not the female mates with a male is decided largely by her.

Despite the larger size of male bonobos, they are not, in contrast to chimps, the dominant sex. Nor are females necessarily dominant. Bonobo society in the wild is more of a mix—what's sometimes called co-dominance. Aggression is infrequent, almost never violent, and when it does occur it has a much more even outcome, with males and females emerging victorious at roughly equal rates. For reasons we don't yet fully understand, bonobos in captivity seem to behave slightly differently from their wild relations. In zoos and parks, females can be particularly feisty, rewarding unwanted male attention with aggression. It's been known for females to bite a male's finger off; in one case, a Bobbitt-like attack occurred when a female severed a male's penis. This kind of savagery seems far less common in the wild, but even so, females do enjoy high status. They have priority at feeding spots, to the point where males will surrender their position to an incoming female. They also get to decide when and where their party moves.

Female bonobos are far more socially inclined than their chimp counterparts and form strong, lasting relationships with one another. This helps if they ever need to present a united front against any flare-ups of male aggression, while at the same time serving another purpose in actually helping the males: although male chimps rely on affiliations with other males to gain status, male bonobos are reliant on their mothers, and the sons of high-ranking females gain benefits from her rank. Indeed, Mum (and even Grandma) may assist when her son conflicts with another male and, rather more oddly from our point of view, mother bonobos take a very direct interest in their sons' sex lives. Introducing their sons to female society helps them get more matings, which can in turn lead to more grandchildren. Essentially, bonobo mums fulfill the role played by other males in chimp society, supporting their adult sons to the extent that male bonobos don't form the brotherhood gangs seen in chimp males. The support of female relatives means that young males can freely interact with adult males without fear of violence.

That's not to say that male bonobos are overly dependent mummy's boys—they can take care of themselves and of one another. Evolutionary biologist Martin Surbeck and evolutionary anthropologist Gottfried Hohmann described the aftermath of an attack by poachers who killed a female bonobo and left her two sons, an infant and another slightly older juvenile, motherless. Females provide all the childcare in bonobos, and at this young age the two sons would have been heavily dependent on her. It was all the more remarkable, then, that a year and a half after they had last been seen, the two males reappeared in their community, the infant riding on his elder

brother's back. They'd survived the trauma and were apparently in-
separable. The extra, unaccustomed child-rearing effort showed in
the older male's threadbare appearance, but that the youngster survived at all is testament to the close bonds between brothers.

Another contrast with chimp society can be seen in the fate of females who move to new communities as they reach maturity. This is
an incredibly dangerous time for chimpanzees, who face an extended
period of adjustment, often accompanied by hostility and violence.
Not so in bonobos. Strangers are welcome in their society, and a new
girl gets a lot of attention, without the accompanying aggression seen
in chimps. Naturally, being bonobos, there's lots of sex, especially
with the resident females, but the immigrants also associate closely
with the dominant females, following them around and begging for
food. They scrounge even when they're surrounded by fruit that they
could quite easily claim for themselves. Lower-ranking apes and newcomers beg from dominants because, it seems, it's about building relationships. Bonobos do not like to dine alone. Two football-sized
rainforest fruits, junglesop and breadfruit, are particularly good for
sharing, and over such feasts immigrants integrate themselves with
the local females and become part of their society.

The peaceful nature of bonobo communities also extends to their
encounters with their neighbors. In chimps, this can be the trigger
for lethal aggression. Bonobos, however, have their own way of dealing with neighboring groups. At first, they may be cautious. When
they hear the calls of strangers in the neighborhood, they may show
intense interest, but they don't necessarily rush to greet them—they
may even move away to avoid them. When two groups do meet, perhaps at the edge of their respective ranges, there's a great deal of calling between them, and though there's some displaying by the males,
fighting is rare. Instead, in true bonobo style, the females take the
lead in mingling, and members of the two groups engage in plenty
of genital rubbing to soothe one another. Eventually the two groups
might merge together and have a feast in a fruit tree. Still, the males
aren't as keen as the females about the proceedings. They often hang

back on their own side, sometimes appearing to try to encourage the females to leave the event, though they don't actually leave until the females are ready. Since the females are enjoying themselves, socializing and having sex with new and interesting partners, this can often take some time: you can't rush these things. It's interesting to note that although the meetings between communities can involve a lot of genital rubbing between females, this isn't necessarily about surrendering to wild, libidinous pleasure. When a female builds a close relationship with another female, the two spend a lot of time together and engage in plenty of mutual grooming, but they don't tend to go in for the genital rubbing so much. This behavior, which seems from the outside to be so overtly sexual, is really much more about meeting, greeting, and reassuring each other in tense situations.

APES LIKE US

For those seeking to understand the genetic basis of human behavior, of our social and cultural norms, apes provide a rich and valuable source of inspiration. Chimpanzees became a particular focus not only because of their close relationship to us, but also because of their large, mixed-sex societies. When early assumptions about chimpanzees as benign, peaceful animals gradually gave way to reports from the wild of their frequent low-level aggression, their sexual conflicts, and their occasional murderous brutality, the picture became more complex. Because of our sense of parallels with the animal, the findings raised disturbing questions about humanity. Some seized on the research to theorize about ruthless competition and aggressive domination among humans. It's in our genes, they said—some even implying that we can't fight it. Others offered more nuanced responses, arguing that our character, like that of the chimps, was a balance of many different tendencies. But the idea of reactive, aggressive chimps as a basis for understanding human nature nonetheless took hold.

And then came a wave—a small wave, admittedly—of studies on bonobos. They are just as closely related to us, and apparently lack the win-at-all-costs mentality that we've regrettably superimposed onto chimps. This is the bonobo paradox: if parts of our behavior are indeed an echo of an ancestral time, then the peaceful bonobo provides just as valid a model as the supposedly warlike chimp. Closer inspection and assessment of our two ape cousins is in order. What can comparisons between chimpanzees and bonobos tell us about ourselves?

Their performance in tests of intelligence tend to conclude that they're at a similarly high level overall. Chimps are superior systematizers, while bonobos are better at empathizing. That is to say, chimps do comparatively well when they have to figure out how things work and how they link together. In the wild, chimps make and use a number of fairly sophisticated tools, while bonobos have relatively few—they use leaves to collect water, or to swat at troublesome insects, but little else. Where bonobos score particularly high is in their social awareness, their ability to read and understand others. One amazing study tracked the gaze direction of chimps and bonobos when they were given pictures of other chimps and bonobos, respectively, to see on which areas of the portraits the two species focused their attention. Chimps looked all around the face of another chimp's portrait, while bonobos lingered on the eyes of an image. In another picture, this time of the full body, chimps looked mostly at the bum, whereas bonobos divided their attention between the face and the bum. When looking at a third picture, of an ape holding an object in its hands, the chimps examined what they were carrying, while the bonobos were interested in both the face and the object. Essentially, the gaze patterns of bonobos followed a similar pattern to that shown by humans in similar experiments, and particularly the patterns shown by people with strong social tendencies.

The continuum between systemizing and empathizing is often applied to humans, with men tending toward the former and women toward the latter. Does this mean that chimps behave more like men and bonobos more like women? Well, no. As interesting as such a comparison might be, it's a gross oversimplification. For one thing, it conveniently forgets the massive overlap between men and women, and between chimps and bonobos. For another, it's a very human trait to try to categorize a continuous group into separate camps. Having said that, there's an interesting and odd pattern that shows itself in chimps, bonobos, and humans. If you compare the length of your second finger, or index finger, with the length of your fourth, the ring finger, what you see depends, to some extent, on your sex. The ring finger is generally a little longer than the index finger, but the typical pattern is that the difference between them is greater in men than in women. This is thought to relate to exposure to androgens—male hormones—before birth. Greater concentrations of androgens produce greater disparity in finger length. If that was all it did, it would be trivial, of course, but it may not be all it does—and the unseen effects on brain development could be important. Bonobos' hands are much like ours in this respect, whereas in chimps the difference can be more pronounced. This suggests the possibility that the brains of unborn chimps are shaped by high levels of testosterone, which in turn might explain some part of their increased aggressiveness in later life.

There are some subtle and potentially important differences between the brains of bonobos and chimps that underscore the differences we see in their behavior. Relative to chimps, bonobos have greater development of and stronger connections between the parts of the brain that relate to how they respond to distress in others; these parts of the brain also govern a primate's capacity to do harm and its general emotional responsiveness. In these characteristics, bonobos are more similar to us than chimps are, though in both, the brain is around a third the size of ours. Profiling the hormones that run through their bodies also helps us to understand the differences in their behavior. When facing conflict, male chimps get a boost of testosterone, priming them for aggression. Bonobos in a similar situation get a rush of cortisol, which can be interpreted as meaning that conflict makes them edgy and nervous. They temper this anxiety using both sex and play. Young chimps are extremely playful, but their sense of fun wears thin as they age, while bonobos carry on larking about even as adults.

Differences between the species can be seen in the ways that they communicate with one another as well. The strict, enforced hierarchy among chimps means that it's necessary for a low-ranking member to give a pant-grunt to acknowledge its subservience to a dominant individual. Bonobos don't seem to have an equivalent; perhaps in their more relaxed society they have no need for kowtowing. When they find food, bonobos produce an array of different sounds according to how appealing they find that food—just as we do. They also call to others when they find a tree laden with fruit, and wait for them before climbing the tree. Chimps, so far as we know, have only one feeding sound, and though they may call out when they make a discovery of food, they usually start in right away, rather than waiting for others to catch up. Both species protest loudly when attacked; bonobos, in particular, seem to save their greatest indignation for occasions when their expectations about what should, and what shouldn't, happen in their society are violated. In other words, it isn't the severity of the aggression that bothers them so much as what we might characterize as its lack of fairness, or even decorum.

Again just like us, bonobos use different communication strategies with different individuals. If a bonobo is addressing a friend, and the message isn't getting across, it will repeat the point in more detail. The two individuals here know one another well, and an understanding exists between them, so repetition with a little clarification works. When the same interaction occurs between bonobos that don't know each other so well, instead of simply repeating the message, and relying on a shared frame of reference, the bonobo will explain it in a different way, using an alternative means of getting the message across.

Even though chimps and bonobos communicate in different ways, both have overlaps in style with us. For instance, we aren't always consciously aware of the structure that occurs in a conversation; what we do notice is when that structure breaks down. When two adult humans are chatting, there's typically a cooperative taking of turns to speak. In conversation with young children, or with a boorish adult, the turn-taking breaks down and people talk over each other, which we naturally find irritating. Both chimps and bonobos show the same subtle back-and-forth dance in their communication—bonobos even prioritize looking at the individual they're addressing, much as we do.

We can distinguish about a dozen different sounds made by each species, and both chimps and bonobos moderate the expression of these extensively, using different pitches and volumes. The crucial thing about human language is how we combine sounds to convey a rich diversity of meanings. English, for example, has about forty-four different phonemes—distinct units of sound—to construct countless different words. Could our ape cousins be doing something similar? We know that bonobos in particular combine their different call types into sequences. Whether this is to convey subtle differences in meanings is unclear, but they certainly pay close attention when listening to these combinations. Studies of communication in another primate, Campbell's monkeys, have looked at this same issue of how they put sounds into sequences and, in doing so, modify their meaning. These monkeys make a variety of calls, including a *krak* call to say there's a leopard around; a slightly modified one, *krak-oo*, to represent a more general concern; and the *boom*, a call used in a social context. But remarkably, the monkeys sometimes combine these sounds to describe something totally different: *boom boom krak-oo krak-oo* apparently means that a tree or a branch is falling. While we don't know whether our ape cousins do this, in their turn-taking and practice of combining sounds into sequences, their conversations do have parallels with ours.

Sound is only one part of ape chat. It's essential for long-range communication in the forested area where they often make their homes. When they're at close quarters, however, they layer body language on top of vocalizations to get their point across. They have, like us, very expressive faces and use dozens of different gestures. But it's interesting to note that they don't use gestures simply as support to the voice—instead, gestures, used in sequence between these apes, often form the basis of a conversation in their own right. The most casual observer can eavesdrop on their gestural repertoire, watching as they negotiate the challenges of their complex social world. It's easy enough to interpret many of them: infants demanding attention from their harassed mothers, or adults begging for food or grooming attention, or soliciting support, or confronting rivals. Their integration of voice, face, and gesture make them expert communicators.

Cooperation is a major part of social life in both bonobos and chimps, but there are some interesting differences between them. In one experiment, a bonobo was placed in a room with a dish of food. In a room to the side, behind a closed door, there was a familiar individual from the same community. The bonobo with the food now had a choice—it could keep its table for one, or it could share by letting the other bonobo in. Slightly surprisingly, most bonobos in this situation decided to be greedy and hog most of the food for themselves. I say surprisingly because, as we've seen, bonobos share readily in the wild and will happily tolerate others joining them at a feast.

When the experimenters ran the test on the bonobos again, however, this time with two bonobos in adjacent rooms—a stranger in one and a familiar bonobo in the other—there was a different outcome: the feeding bonobo let the stranger in to share, and the stranger, in turn, let the other bonobo in, so that all three shared the food. By inviting strangers to lunch, they lost food but made a new friend. It's not an experiment that could be done with chimps— they're much less keen on strangers altogether, and the outcome might be violent. Yet, despite their reluctance to share in some contexts, chimps and bonobos will partition food when the whole group is present, even when it's possible for one individual to keep it for himself or herself. The main difference comes in the way that newcomers try to get access to the food: chimps beg for it, whereas bonobos, naturally enough, offer sex.

Experiments examining sharing and cooperative behavior in these two complex animals produce mixed results: chimps, in particular, sometimes share and sometimes don't. Both species seem to be rather more cooperative in the wild than in captivity. We'd expect, then, that in an experiment on captive chimps where access to a food treat got progressively more difficult over time, they'd respond by getting more feisty with each other. This hypothesis was tested in a group of chimps at Lincoln Park Zoo in Chicago using an artificial termite mound baited with ketchup. The chimps learned to dip sticks into holes in the mound to get the tasty sauce. To start with, there were enough holes for each chimp to be able to have one to itself. This number was then reduced one by one, so that there was eventually a shortage of the previously abundant food source. (It's important to point out that this wasn't the only food source for the chimps, but a highly desirable extra.) In contrast to expectations, rather than increasing the competition, the shortage caused the chimps to react with patience and sharing: they waited their turn to dip for food. That's the thing about these intelligent and complex animals—they constantly surprise us.

Making generalizations about the behavior of any species is tricky—there's always an exception to prove the rule, and this is never more the case than with apes. For instance, there's much more violence between communities of East African chimpanzees than among their West African counterparts. There are plenty of potential reasons underlying this difference. Just as history provides us with examples of both peaceful and warlike human societies, there may be an element of culture to it. Social groups tend to enforce patterns of behavior on their members, so that they conform to a norm. Primates, like other animals, like to fit in. Rhesus monkeys are among the more aggressive of macaque species, yet, if you take a youngster and slot it into a group of the more easygoing stump-tail macaques, it adopts their rules, becoming more easygoing itself. Similarly, hamadryas baboons and olive baboons, who are at opposite extremes of a continuum of baboon aggression, will modify their behavior if they're transplanted into each other's groups. They, too, adapt to fit in.

The power of the social environment in shaping long-term behavior can be seen in the cultural conformity of a troop of baboons in Kenya. They lived near a local garbage dump, where the allure of free food was tempered by the need to run a gauntlet of rival baboons and angry humans. Consequently, only the ballsiest, most belligerent males in the troop took to raiding the garbage dump. Their foraging for free, though foul, food was ultimately the end of them—those aggressive males were wiped out by tuberculosis.

The remaining baboons of the troop—the females and the more relaxed males—now found themselves living in a much more harmonious society. You'd imagine that this would last only as long as it took aggressive immigrant males to move in and take over the troop, but twenty years later, this still hadn't happened. The surviving males had by then died, and the males that had joined the troop were playing by its unusual rules—a less rigid approach to the hierarchy, coupled, naturally enough, with plenty of grooming, had become established in this exceptionally peaceful troop of baboons. Can culture explain differences between communities of chimpanzees? While it's hard to categorically state that it's the greatest influence on chimp behavior, it does play a role. The garbage dump baboons highlight a pertinent lesson: the difference between violence and peacefulness is often just a matter of conformity.

We share aspects of our behavior with both of our closest ape cousins. In some cases, we're more like chimps; in others, more like bonobos. It's impossible to know at this remove which characteristics each of us share with our long-extinct shared ancestor, who perhaps had a mix of the traits now seen in our three species. Examinations of our living, modern-day ape relatives help us as we seek answers to the questions about the origins of human nature. Even now, six million years since we split from them, many aspects of the behavior of chimps and bonobos are instantly recognizable to us.

EPILOGUE

I am, because we are.

—**Ubuntu philosophy**

Sociality is a fundamental part of our existence. Our lives are interwoven with those of our friends and our families. Our societies are structured according to relationships that provide the basis for our economies and our governments. They are the foundation for our culture, for the development of human civilization, and, ultimately, for the success of our species. As we've seen in this book, we are far from alone in this tendency to be social. Indeed, the transition from a lone existence to living in groups was one of the most important evolutionary developments in the history of life on Earth.

We can measure just how significant living with others is in many different ways. The consequences for being deprived of access to social interactions can be severe for group-living animals. For instance, if a herring is taken from its school and kept on its own for any amount of time, it succumbs to the stress of its exile—lending credence, perhaps, to the phrase "dying to meet you." Essentially, it dies of loneliness. We may pity a moribund fish and think we're not like them, yet solitary confinement remains one of the most feared punishments in

327

the justice system. The prolonged isolation induces depression, even hallucinations; divorced from contact with other people, the human mind starts to collapse in on itself. The flip side is that our social relationships have a powerful influence in promoting mental health and longevity. The effect is incredibly powerful: engagement with a wide circle of friends and the maintenance of fruitful relationships is an even stronger predictor of living to a ripe old age than physical exercise. Again, this isn't just a human thing; we see something similar in baboons, rats, and crows, among many others. The support that an individual gains through secure social bonds, and the buffering from the vicissitudes of life that takes place within a social group, provide impressive benefits to health and well-being.

The natural world is full of wonderful examples of the importance of sociality. We can see it in a flock of pigeons detecting a stooping raptor and taking evasive action in time for all to live another day, or in the pulses of information shooting through a shoal of fish as they turn to avoid the strike of a predator. We see it in ant foraging networks, as each little six-legged Theseus lays down a trail and they all collectively work out the best route to a dropped ice cream cone, and we see it in the dance of bees guiding their sisters to a newly opened flower rich with nectar. We can marvel, too, at the harrying of a pack of wolves, or at the teamwork of orcas combining to create a big enough wave to wash a seal from an ice floe. We can even smile at a posse of Harris's hawks, who, looking for their next meal in arid landscapes with few perches, stand on each other's backs, like a living totem pole, so that the topmost bird can gain a better view.

The rewards of company in our day-to-day lives are one thing, but if we delve back into our evolutionary history, we can see how sociality has shaped us and other animals in a fundamental way. The reasons that animals congregate vary from species to species, but there are many common factors. Groups provide a refuge against predators, and they allow access to information about, for instance, where to find the next meal. In this setting, individuals increase their chances of survival and then raise more offspring. Where the youngsters are raised in a cooperative group, infants and juveniles interact together; they develop their skills, they become socialized. As animals progressed to living in groups, they underwent changes, interacting in a more sophisticated fashion, even cooperating with fellow group members, and became able to achieve more together than they could alone. Social behavior developed, and cultures evolved.

The transition to social living even changed the genetic architecture and biochemistry of individuals. The tendency of animals to seek out others is written into their DNA—the sociability of zebrafish and the preference of birds to join larger breeding colonies are shaped by genes. When getting together in groups is advantageous, natural selection acts to promote this behavior, so the genes that give an animal the predisposition to seek out others are passed on to the next generations. To some extent, the same is true for us. The connections that we establish within our own social networks is, in part, determined by our genes. In particular, the extent to which people warm to us and name us as their friend has a strong genetic component, as does the likelihood that we will operate within a strong social clique. Genes regulating the expression of hormones that affect personality traits such as agreeableness and extraversion, including oxytocin, may play a large role in the inherited disposition to be sociable.

We are chemical creatures. How two animals respond to one another when they meet is influenced by hormones. Their responses might be anywhere on a continuum between aggression and affiliation. Comparisons between species show how this works. Those species that show territorial behavior are more inclined to be aggressive when an outsider of their own kind shows up; they repel one another like mismatched magnets. If social animals were uniformly aggressive toward one another, then groups simply couldn't form. The basis of grouping is social attraction—flip the magnets around and they glom together. Within the brains of these species, hormones come into play to adjust and fine-tune this circuitry. In the specific parts of the brain that govern how animals interact with one another, sociable animals have more cells producing and responding to specific hormones than animals that go their own way. How animals respond to stress is also shaped by hormones. Social animals respond less dramatically to stressors in the presence of their own kind: this is the so-called social buffering described earlier. Hormones even affect who we associate with—they can make us predisposed to affiliate with friends and relations, or with strangers and non-relatives. Changes in hormonal levels over time and in response to different situations allow animals to adapt their responses to one another—sometimes they are more affiliative, sometimes a bit more Greta Garbo–like in wanting to be alone.

And then there's how animals talk to one another. Although all animals communicate, those that live in groups need an extended vocabulary of signals to allow them to interact effectively. The functioning of these groups relies on their communication; without it the group can't coordinate its activities or understand the relationships within their society. Imagine two different species of animals, one that lives in a territory with its mate and one that lives in a social group. The first animal has only to whisper sweet nothings to its mate and to swear from time to time at its neighbors whenever they encroach. The animal that lives in a group has to be a little more eloquent, interacting daily with a host of other group members. It has to be able to distinguish between rivals and affiliates, and it has to make sure that it, in turn, can be identified. It must be able to make its motivations clear, navigate relationships within a hierarchy, and adjust its responses to others depending on their respective social ranks. The pressure on animals to make themselves understood leads to differences between species according to whether they are loners or sociably inclined. We even get differences within species—chickadees, small North American birds named for their chattering call, adjust the complexity of their calling to the group size they're in. Language coevolved with sociality; it was the need to negotiate and coordinate behavior with a group that drove the development of language in humans, and the reason I can communicate with you now through this book.

The study of animal communication has come a long way in recent times. For decades, we tried to school animals like children, concentrating our effort on coaching them to express themselves using sign language or pictures. While the linguistic accomplishments of animals such as Alex the African gray parrot or the host of apes that have been painstakingly taught over the years are impressive, they tell us little about natural communication between animals. Language, and ultimately communication, is about a whole lot more than learning "words," however they're represented. Human conversation works on the basis of shared experience, an understanding of a general frame of reference. We have shaped it and, in turn, it has profoundly shaped us. Although the motivation to get a sense of what and how an animal thinks and feels is understandable, it doesn't make much sense (to me, at least) to do so by teaching them a human-centric scheme. A much better method is to study how they communicate with each other. The problem is that it's a hell of a language barrier. But we're making progress.

Animals communicate in a wonderful variety of ways. Take what happens when we greet each other. Social animals have a marvelous and diverse array of approaches to greetings. In parts of Yorkshire where I grew up, especially among farmers, a barely perceptible nod was a sign of high favor, to be delivered only on the basis of a minimum twenty-year acquaintanceship. Less restrained people might shake hands, bump fists, or, if they're French, descend into a terrible amount of kissing. I was witness to a characteristically excitable greeting between two teenaged girls in a large public square in Berlin. The pair converged at a pace from a distance, squealing as they ran, arms spread. Alas, since they were of rather different sizes, they failed to account for the rather large difference in momentum, and when they met, the portlier of the two leveled the other with an accidental ferocity that would have seen her sent off in a rugby match. Yet despite the multifarious ways that we greet friends, we can't match the diversity seen in the animal kingdom. Lobsters urinate in each other's faces as a "how d'you do," while dogs, of course, are inveterate bum sniffers. Cichlid fish buzz at the return to the nest of a partner. White-faced capuchins say hello by sticking their fingers up their chum's nose, while the same message in male Guinea baboons is achieved by fiddling with a friend's phallus. It's a demonstration of trust, you see, though, on balance, I prefer the Yorkshire approach.

Another window into group communication is provided by the behavior of animals when they want to move. The trouble here for group-living animals is when they try to balance their own preference to go in a particular direction with their need to remain with the group. No one wants to go it alone, but they don't want to pass up the chance to act on their own information, either. The result can be a kind of deadlock, one that I've experienced myself only too many times. Get a group of more than about three people together and try to work out where to go for dinner. It's agony. When coral reef fish move between coral heads, you can watch as the pre-departure activity builds within the group, like a pan of water coming to the boil. One of them will dart in the direction that it wants to go, but if no one follows, it puts the brakes on and retreats, stymied, back to the group. If it's determined, it will keep feinting like this until it recruits enough followers to make the crossing.

Sometimes animals make do without all this preamble and agitation. In which case, how do they work out who's got the best information? Sometimes they can read it in the behavior of the chooser. If it has good information, it will move with confidence, which is to say, in a clear direction, with little hesitation or meandering. Such a clear and forthright signal is likely to attract followers. At other times, the most dominant animal unilaterally decides and the rest simply have to fall in behind it. African wild dogs usually have to follow a route chosen by the leader of the pack, but, just occasionally, a lesser individual points its body in the direction it wishes to go and then sneezes. Other pack members might pick up on this and join in, sneezing to show their support. If enough of them get together to support the move, the sneezers win out over the dominant member and the pack sets off snottily on their new course.

African buffaloes live in huge herds that roam and rest together. Following a period of slumbering, the herd has to decide where to go, and they do this by voting. Only the females get a say—the juveniles have to fall in with Mum, and the males have to abide by the decision if they want to stay with them. Each cow stands up and gazes in her preferred direction. Once the votes are in, the herd moves off in the direction the majority chose.

Something similar happens in Tonkean macaques. When a troop of these little monkeys is ready to make a move, an individual will take a few paces along its favored route and then pause to look back at the rest. Then the next macaque votes. It might go along with the first, or it might suggest a different direction. Subsequently, each monkey lines up behind whichever candidate and course it prefers until they have a majority, at which point the losing team gives up and gets behind the majority preference, and the group moves off. When gorillas move, it's usually with the old silverback, the dominant male, leading the way. He might fondly imagine himself to have decided on the group move, but the issue was more than likely decided by the females in his group earlier. They rouse the gorilla gang to get ready to move and even decide the direction, until all that's left for the silverback to do is to step out, apparently leading, but, in reality, simply lending his authority to a decision that's already been made.

Among all animals, one group in particular—the primates—is exceptional for its members' large brains. The question is, why? There are lots of possible explanations: perhaps it's the rich diet of fruit that fuels the prodigious development of their gray matter, or maybe the need to build a mental map as they roam far and wide in their forest home. Or perhaps it's their sociability. Of the many different species of primates alive today, most are social to some extent, but we see a wide range of different group sizes, from those who live with a limited circle of friends to those that are exuberantly gregarious and live in large communities.

Back in the early 1990s, the evolutionary psychologist Robin Dunbar set out to examine these different factors. His findings were clear—the single most important driver of big brains was the group size in which the animals lived. In particular, it was the neocortex—the most advanced part of the brain, the area associated with cognition, detailed sensory perception, reasoning, and communication—that was most strongly predicted by group size. Being part of a society is challenging for primates. They must negotiate a potential minefield of dynamic, shifting relationships. To be successful, they have to recognize individuals and understand how they relate to them, then they must tailor their behavior accordingly. Collecting and processing this mass of social information and deciding how to use it places high demands on the cognitive ability of group members. It's vital for social animals to have the intelligence to navigate the interactions and intrigues within their groups. As group size increases, the number of interactions to keep track of grows exponentially. For all this, you need significant cognitive capacity, and for many species, the larger the group, the larger the brain must be.

Of course, group size is only one facet of sociality. Lots of fish form gigantic shoals—if it was just about the number of individuals in a group, you'd expect fish to be winning Nobel Prizes. The fact that you've never read a novel by a goldfish isn't just because they struggle with typing. The nature and complexity of social relationships is as important, arguably more important, than the basic quantity. Schooling fish are highly responsive to the changes in the behavior of near neighbors, but they don't form enduring associations with one another. Big brains are only an advantage when animals remain together over an extended period of time, participating in a stable social group of individuals with whom they interact frequently, and whose characteristics and peculiarities they learn. As we've already seen, animal societies can be cloak-and-dagger affairs, featuring plenty of maneuvering and politicking, and an animal must be adept at building social alliances if it is to gain influence and power. For those who intrigue and connive to reach the higher echelons of their society, success is dependent on sophisticated cognitive skills and rewarded through increased reproductive opportunities.

We might imagine a large brain to be essential, but it's an expensive luxury. Maintaining this organ, which is just 2 percent of our weight, uses 20 percent of our energy intake. Your brain devours the fuel in two Mars bars every single day, using roughly the same amount of power as the leg muscles of an athlete running a marathon, and it never stops. Natural selection doesn't tend to equip animals with unnecessary adaptations. A snail with a giant brain might be a philosophical marvel, a genius of the garden, but it wouldn't be better at being a snail. In fact, the energy it would use up powering its slimy intellect would detract from its other activities, including having baby snails. Ultimately, a hyperintelligent snail would be a worse snail.

Nature is parsimonious like that. Plenty of animals get along just fine with only a modicum of gray matter, just enough for their own behavioral repertoire. Snails are one such, and, simple though they are, they're an incredibly successful group. For animals that live in complex groups, characterized by convoluted, interwoven relationships, the kind of social intelligence that can only be realized by having a large and complex brain is essential. That's why, if you compare measures of brain size against group size among similar animals, you can see a pattern. In primates, the largest brains tend to belong to those that form the largest groups. The same is true of bats, cetaceans, carnivores, ungulates, even ants and wasps. We can trace the growth of brain size in the fossil record. Over millions of years, the brains of social mammals in particular have shown a steady but relentlessly increasing trend.

Group living isn't the only path to intelligence. Not all social animals have giant brains, and not all animals with giant brains are social. Bees and other insects have tiny brains, despite the highly organized social lives that many of them enjoy, yet they have phenomenal spatial memory, are capable of some impressive feats of learning, and can build complex homes. Bees can even be pessimistic—a case of the nectary being half empty. Moreover, some social insects actually have less brain tissue than their solitary counterparts, relying on collective cognition rather than individual brilliance. By the same token, nutcrackers aren't especially social birds, but they have impressive brains. For them, it's a matter of remembering where they've hidden seeds to see them through the winter. They might hide as many as a hundred thousand seeds each autumn and can remember where they've put them even months later—bear that in mind next time you lose your keys. Nonetheless, it seems likely that our human brains, which allow us to contemplate the meaning of life, the universe, and everything in it, were fostered by the need to navigate our place in our ancient societies.

If social living propelled the development of specialized cognitive skills and larger brains, the process opened the door to all manner of other intellectual developments. The behavioral flexibility that permits problem-solving and innovation is a characteristic of intelligent, large-brained species and one that propels their adaptability and success. Alongside this quality of innovation is the tendency of social animals to emulate and imitate one another, so that what one learns, all learn. This social learning provides a powerful means for know-how and knowledge to spread, leading to traditions within groups and, ultimately, the evolution of culture. We know that large-brained animals are more likely than small-brained ones both to innovate and to learn from one another, so it may be that the cognitive skills required for group living developed in tandem with innovation and cultural transmission, augmenting the benefits of large brains and propelling their further development.

The animals that have disproportionately large brains for their size—including chimpanzees, dolphins, elephants, and, of course, humans—are all social. This being the case, we might borrow from Monty Python and ask what, apart from intelligence, language, longevity, consciousness, reasoning, social learning, and culture, has sociality ever done for us?

FURTHER READING AND SELECTED REFERENCES

CHAPTER 1: BROWN ALE AND CANNIBALISM

Coyle, K. O., and A. I. Pinchuk. (2002). The abundance and distribution of euphausiids and zero-age pollock on the inner shelf of the southeast Bering Sea near the Inner Front in 1997–1999. *Deep Sea Research Part II: Topical Studies in Oceanography*, *49*(26), 6009–6030.

Willis, J. (2014). Whales maintained a high abundance of krill; both are ecosystem engineers in the Southern Ocean. *Marine Ecology Progress Series*, *513*, 51–69.

Tarling, G. A., and S. E. Thorpe. (2017). Oceanic swarms of Antarctic krill perform satiation sinking. *Proceedings of the Royal Society B: Biological Sciences*, *284*(1869), 20172015.

Margesin, R., and F. Schinner. (1999). *Biotechnological applications of cold-adapted organisms*. Springer Science and Business Media.

Everson, I. (ed.). (2008). *Krill: biology, ecology and fisheries*. John Wiley and Sons.

Fornbacke, M., and M. Clarsund. (2013). Cold-adapted proteases as an emerging class of therapeutics. *Infectious Diseases and Therapy*, *2*(1), 15–26.

Kawaguchi, S., R. Kilpatrick, L. Roberts, R. A. King, and S. Nicol. (2011). Ocean-bottom krill sex. *Journal of Plankton Research*, *33*(7), 1134–1138.

Rogers, S. M., T. Matheson, E. Despland, T. Dodgson, M. Burrows, and S. J. Simpson. (2003). Mechanosensory-induced behavioural gregarization in the desert locust *Schistocerca gregaria*. *Journal of Experimental Biology*, *206*(22), 3991–4002.

Simpson, S. J., G. A. Sword, P. D. Lorch, and I. D. Couzin. (2006). Cannibal crickets on a forced march for protein and salt. *Proceedings of the National Academy of Sciences*, *103*(11), 4152–4156.

Lihoreau, M., L. Brepson, and C. Rivault. (2009). The weight of the clan: even in insects, social isolation can induce a behavioural syndrome. *Behavioural Processes*, *82*(1), 81–84.

CHAPTER 2: HONEY, I FED THE KIDS (AND NOW I'M GOING TO EXPLODE)

Wcislo, W., and J. H. Fewell. (2017). Sociality in bees. In D. R. Rubenstein and P. Abbot (eds.), *Comparative social evolution*, 50–83. Cambridge University Press.

McDonnell, C. M., C. Alaux, H. Parrinello, J. P. Desvignes, D. Crauser, E. Durbesson, B. Dominique, and Y. le Conte. (2013). Ecto- and endoparasite induce similar chemical and brain neurogenomic responses in the honey bee (*Apis mellifera*). *BMC Ecology*, *13*(1), 1–15.

Watanabe, D., H. Gotoh, T. Miura, and K. Maekawa. (2014). Social interactions affecting caste development through physiological actions in termites. *Frontiers in Physiology*, *5*, 127.

Wen, X. L., P. Wen, C. A. Dahlsjö, D. Sillam-Dussès, and J. Šobotník. (2017). Breaking the cipher: ant eavesdropping on the variational trail pheromone of its termite prey. *Proceedings of the Royal Society B: Biological Sciences*, *284*(1853), 20170121.

Oberst, S., G. Bann, J. C. Lai, and T. A. Evans. (2017). Cryptic termites avoid predatory ants by eavesdropping on vibrational cues from their footsteps. *Ecology Letters*, *20*(2), 212–221.

Röhrig, A., W. H. Kirchner, and R. H. Leuthold. (1999). Vibrational alarm communication in the African fungus-growing termite genus *Macrotermes* (*Isoptera, Termitidae*). *Insectes sociaux*, *46*(1), 71–77.

Yanagihara, S., W. Suehiro, Y. Mitaka, and K. Matsuura. (2018). Age-based soldier polyethism: old termite soldiers take more risks than young soldiers. *Biology Letters*, *14*(3), 20180025.

Šobotník, J., T. Bourguignon, R. Hanus, Z. Demianová, J. Pytelková, M. Mareš, P. Foltynova, et al. (2012). Explosive backpacks in old termite workers. *Science*, *337*(6093), 436.

Rettenmeyer, C. W., M. E. Rettenmeyer, J. Joseph, and S. M. Berghoff. (2011). The largest animal association centered on one species: the army ant *Eciton burchellii* and its more than 300 associates. *Insectes sociaux*, *58*(3), 281–292.

Kronauer, D. J. C., E. R. Ponce, J. E. Lattke, and J. J. Boomsma. (2007). Six weeks in the life of a reproducing army ant colony: male parentage and colony behaviour. *Insectes sociaux*, *54*(2), 118–123.

Franks, N. R., and B. Hölldobler. (1987). Sexual competition during colony reproduction in army ants. *Biological Journal of the Linnean Society*, *30*(3), 229–243.

Mlot, N. J., C. A. Tovey, and D. L. Hu. (2011). Fire ants self-assemble into waterproof rafts to survive floods. *Proceedings of the National Academy of Sciences*, *108*(19), 7669–7673.

Deslippe, R. (2010). Social parasitism in ants. *Nature Education Knowledge 3*(10), 27.

Brandt, M., J. Heinze, T. Schmitt, and S. Foitzik. (2006). Convergent evolution of the Dufour's gland secretion as a propaganda substance in the slave-making ant genera *Protomognathus* and *Harpagoxenus*. *Insectes sociaux*, *53*(3), 291–299.

Seifert, B., I. Kleeberg, B. Feldmeyer, T. Pamminger, E. Jongepier, and S. Foitzik. (2014). *Temnothorax pilagens* sp. n.—a new slave-making species of the tribe *Formicoxenini* from North America (*Hymenoptera, Formicidae*). *ZooKeys*, *368*, 65.

Jongepier, E., and S. Foitzik. (2016). Ant recognition cue diversity is higher in the presence of slavemaker ants. *Behavioral Ecology*, *27*(1), 304–311.

Zoebelein, G. (1956). Der Honigtau als Nahrung der Insekten: Teil I. *Zeitschrift für angewandte Entomologie, 38*(4), 369–416. (cited in AntWiki)

Oliver, T. H., A. Mashanova, S. R. Leather, J. M. Cook, and V. A. Jansen. (2007). Ant semiochemicals limit apterous aphid dispersal. *Proceedings of the Royal Society B: Biological Sciences, 274*(1629), 3127–3131.

Charbonneau, D., and A. Dornhaus. (2015). Workers "specialized" on inactivity: behavioral consistency of inactive workers and their role in task allocation. *Behavioral Ecology and Sociobiology, 69*(9), 1459–1472.

CHAPTER 3: FROM DITCHES TO DECISIONS

Kelly, J. (2019). The role of the preoptic area in social interaction in zebrafish (PhD diss., Liverpool John Moores University).

McHenry, J. A., J. M. Otis, M. A. Rossi, J. E. Robinson, O. Kosyk, N. W. Miller, Z. A. McElligott, A. Budygin, D. R. Rubinow, and G. D. Stuber. (2017). Hormonal gain control of a medial preoptic area social reward circuit. *Nature Neuroscience, 20*(3), 449–458.

Couzin, I. D., J. Krause, N. R. Franks, and S. A. Levin. (2005). Effective leadership and decision-making in animal groups on the move. *Nature, 433*(7025), 513–516.

Ward, A. J., D. J. Sumpter, I. D. Couzin, P. J. Hart, and J. Krause. (2008). Quorum decision-making facilitates information transfer in fish shoals. *Proceedings of the National Academy of Sciences,* 105(19), 6948–6953.

Sumpter, D. J., J. Krause, R. James, I. D. Couzin, and A. J. Ward. (2008). Consensus decision making by fish. *Current Biology, 18*(22), 1773–1777.

CHAPTER 4: CLUSTERFLOCKS

Goodenough, A. E., N. Little, W. S. Carpenter, and A. G. Hart. (2017). Birds of a feather flock together: insights into starling murmuration behaviour revealed using citizen science. *PLOS One, 12*(6), e0179277.

Young, G. F., L. Scardovi, A. Cavagna, I. Giardina, and N. E. Leonard. (2013). Starling flock networks manage uncertainty in consensus at low cost. *PLOS Computational Biology, 9*(1), e1002894.

Portugal, S. J., T. Y. Hubel, J. Fritz, S. Heese, D. Trobe, B. Voelkl, S. Hailes, A. M. Wilson, and J. R. Usherwood. (2014). Upwash exploitation and downwash avoidance by flap phasing in ibis formation flight. *Nature, 505*(7483), 399–402.

Nagy, M., I. D. Couzin, W. Fiedler, M. Wikelski, and A. Flack. (2018). Synchronization, coordination and collective sensing during thermalling flight of freely migrating white storks. *Philosophical Transactions of the Royal Society B: Biological Sciences, 373*(1746), 20170011.

Simons, A. M. (2004). Many wrongs: the advantage of group navigation. *Trends in Ecology and Evolution, 19*(9), 453–455.

Dell'Ariccia, G., G. Dell'Omo, D. P. Wolfer, and H. P. Lipp. (2008). Flock flying improves pigeons' homing: GPS track analysis of individual flyers versus small groups. *Animal Behaviour, 76*(4), 1165–1172.

Aplin, L. M., D. R. Farine, J. Morand-Ferron, A. Cockburn, A. Thornton, and B. C. Sheldon. (2015). Experimentally induced innovations lead to persistent culture via conformity in wild birds. *Nature, 518*(7540), 538–541.

Kenward, B., C. Rutz, A. A. Weir, and A. Kacelnik. (2006). Development of tool use in New Caledonian crows: inherited action patterns and social influences. *Animal Behaviour*, *72*(6), 1329–1343.

Grecian, W. J., J. V. Lane, T. Michelot, H. M. Wade, and K. Hamer. (2018). Understanding the ontogeny of foraging behaviour: insights from combining marine predator bio-logging with satellite-derived oceanography in hidden Markov models. *Journal of the Royal Society Interface*, *15*(143), 20180084.

van Dijk, R. E., J. C. Kaden, A. Argüelles-Ticó, L. M. Beltran, M. Paquet, R. Covas, C. Doutrelant, and B. J. Hatchwell. (2013). The thermoregulatory benefits of the communal nest of sociable weavers *Philetairus socius* are spatially structured within nests. *Journal of Avian Biology*, *44*(2), 102–110.

Laughlin, A. J., D. R. Sheldon, D. W. Winkler, and C. M. Taylor. (2014). Behavioral drivers of communal roosting in a songbird: a combined theoretical and empirical approach. *Behavioral Ecology*, *25*(4), 734–743.

Hatchwell, B. J., S. P. Sharp, M. Simeoni, and A. McGowan. (2009). Factors influencing overnight loss of body mass in the communal roosts of a social bird. *Functional Ecology*, *23*(2), 367–372.

Mumme, R. L. (1992). Do helpers increase reproductive success? *Behavioral Ecology and Sociobiology*, *31*(5), 319–328.

Emlen, S. T., and P. H. Wrege. (1992). Parent–offspring conflict and the recruitment of helpers among bee-eaters. *Nature*, *356*(6367), 331–333.

McDonald, P. G., and J. Wright. (2011). Bell miner provisioning calls are more similar among relatives and are used by helpers at the nest to bias their effort towards kin. *Proceedings of the Royal Society B: Biological Sciences*, *278*(1723), 3403–3411.

Braun, A., and T. Bugnyar. (2012). Social bonds and rank acquisition in raven non-breeder aggregations. *Animal Behaviour*, *84*(6), 1507–1515.

Heinrich, B., and J. Marzluff. (1995). Why ravens share. *American Scientist*, *83*(4), 342–349.

Heinrich, B. (1988). Winter foraging at carcasses by three sympatric corvids, with emphasis on recruitment by the raven, *Corvus corax*. *Behavioral Ecology and Sociobiology*, *23*(3), 141–156.

Marzluff, J. M., and R. P. Balda. (2010). *The pinyon jay: behavioral ecology of a colonial and cooperative corvid*. A&C Black.

Bond, A. B., A. C. Kamil, and R. P. Balda. (2004). Pinyon jays use transitive inference to predict social dominance. *Nature*, *430*(7001), 778–781.

Duque, J. F., W. Leichner, H. Ahmann, and J. R. Stevens. (2018). Mesotocin influences pinyon jay prosociality. *Biology Letters*, *14*(4), 20180105.

CHAPTER 5: GETTING INTO MISCHIEF

Feng, A. Y., and C. G. Himsworth. (2014). The secret life of the city rat: a review of the ecology of urban Norway and black rats (*Rattus norvegicus* and *Rattus rattus*). *Urban Ecosystems*, *17*(1), 149–162.

Clark, B. R., and E. O. Price. (1981). Sexual maturation and fecundity of wild and domestic Norway rats (*Rattus norvegicus*). *Reproduction*, *63*(1), 215–220.

Galef, B. G. (1980). Diving for food: analysis of a possible case of social learning in wild rats (*Rattus norvegicus*). *Journal of Comparative and Physiological Psychology, 94*(3), 416.

Hepper, P. G. (1988). Adaptive fetal learning: prenatal exposure to garlic affects postnatal preferences. *Animal Behaviour, 36*(3), 935–936.

Mennella, J. A., and G. K. Beauchamp. (2005). Understanding the origin of flavor preferences. *Chemical Senses, 30*(suppl. 1), i242–i243.

Noble, J., P. M. Todd, and E. Tucif. (2001). Explaining social learning of food preferences without aversions: an evolutionary simulation model of Norway rats. *Proceedings of the Royal Society of London. Series B: Biological Sciences, 268*(1463), 141–149.

Calhoun, J. B. (1973). Death squared: the explosive growth and demise of a mouse population. *Journal of the Royal Society of Medicine 66*(1), 80–88.

Rutte, C., and M. Taborsky. (2007). Generalized reciprocity in rats. *PLOS Biology, 5*(7), e196.

Dolivo, V., and M. Taborsky. (2015). Norway rats reciprocate help according to the quality of help they received. *Biology Letters, 11*(2), 20140959.

Schweinfurth, M. K., and M. Taborsky. (2018). Relatedness decreases and reciprocity increases cooperation in Norway rats. *Proceedings of the Royal Society B: Biological Sciences, 285*(1874), 20180035.

Schweinfurth, M. K., and M. Taborsky. (2018). Reciprocal trading of different commodities in Norway rats. *Current Biology, 28*(4), 594–599.

Stieger, B., M. K. Schweinfurth, and M. Taborsky. (2017). Reciprocal allogrooming among unrelated Norway rats (*Rattus norvegicus*) is affected by previously received cooperative, affiliative and aggressive behaviours. *Behavioral Ecology and Sociobiology, 71*(12), 1–12.

Weaver, I. C., N. Cervoni, F. A. Champagne, A. C. D'Alessio, S. Sharma, J. R. Seckl, S. Dymov, M. Szyf, and M. J. Meaney. (2004). Epigenetic programming by maternal behavior. *Nature Neuroscience, 7*(8), 847–854.

Lester, B. M., E. Conradt, L. L. LaGasse, E. Z. Tronick, J. F. Padbury, and C. J. Marsit. (2018). Epigenetic programming by maternal behavior in the human infant. *Pediatrics, 142*(4), e20171890.

Ackerl, K., M. Atzmueller, and K. Grammer. (2002). The scent of fear. *Neuroendocrinology Letters, 23*(2), 79–84.

Kiyokawa, Y. (2015). Social odors: alarm pheromones and social buffering. In M. Wöhr and S. Krach (eds.), *Social behavior from rodents to humans*, 47–65. Springer.

Gunnar, M. R. (2017). Social buffering of stress in development: a career perspective. *Perspectives on Psychological Science, 12*(3), 355–373.

Morozov, A., and W. Ito. (2019). Social modulation of fear: facilitation vs buffering. *Genes, Brain and Behavior, 18*(1), e12491.

Sato, N., L. Tan, K. Tate, and M. Okada. (2015). Rats demonstrate helping behavior toward a soaked conspecific. *Animal Cognition, 18*(5), 1039–1047.

Ben-Ami Bartal, I., H. Shan, N. M. Molasky, T. M. Murray, J. Z. Williams, J. Decety, and P. Mason. (2016). Anxiolytic treatment impairs helping behavior in rats. *Frontiers in Psychology, 7*, 850.

Muroy, S. E., K. L. Long, D. Kaufer, and E. D. Kirby. (2016). Moderate stress-induced social bonding and oxytocin signaling are disrupted by predator odor in male rats. *Neuropsychopharmacology, 41*(8), 2160–2170.

Pittet, F., J. A. Babb, L. Carini, and B. C. Nephew. (2017). Chronic social instability in adult female rats alters social behavior, maternal aggression and offspring development. *Developmental Psychobiology, 59*(3), 291–302.

Holmes, M. M., G. J. Rosen, C. L. Jordan, G. J. de Vries, B. D. Goldman, and N. G. Forger. (2007). Social control of brain morphology in a eusocial mammal. *Proceedings of the National Academy of Sciences, 104*(25), 10548–10552.

Braude, S. (2000). Dispersal and new colony formation in wild naked mole-rats: evidence against inbreeding as the system of mating. *Behavioral Ecology, 11*(1), 7–12.

CHAPTER 6: FOLLOWING THE HERD

Pitt, D., N. Sevane, E. L. Nicolazzi, D. E. MacHugh, S. D. Park, L. Colli, R. Martinez, M. W. Bruford, and P. Orozco-terWengel. (2019). Domestication of cattle: two or three events? *Evolutionary Applications, 12*(1), 123–136.

Bollongino, R., J. Burger, A. Powell, M. Mashkour, J. D. Vigne, and M. G. Thomas. (2012). Modern taurine cattle descended from small number of Near-Eastern founders. *Molecular Biology and Evolution, 29*(9), 2101–2104.

MacHugh, D. E., G. Larson, and L. Orlando. (2017). Taming the past: ancient DNA and the study of animal domestication. *Annual Review of Animal Biosciences, 5*, 329–351.

Hemmer, H. (1990). *Domestication: the decline of environmental appreciation.* Cambridge University Press.

Ballarin, C., M. Povinelli, A. Granato, M. Panin, L. Corain, A. Peruffo, and B. Cozzi. (2016). The brain of the domestic *Bos taurus*: weight, encephalization and cerebellar quotients, and comparison with other domestic and wild Cetartiodactyla. *PLOS One, 11*(4), e0154580.

Minervini, S., G. Accogli, A. Pirone, J. M. Graïc, B. Cozzi, and S. Desantis. (2016). Brain mass and encephalization quotients in the domestic industrial pig (*Sus scrofa*). *PLOS One, 11*(6), e0157378.

Burns, J. G., A. Saravanan, and F. H. Rodd. (2009). Rearing environment affects the brain size of guppies: lab-reared guppies have smaller brains than wild-caught guppies. *Ethology, 115*(2), 122–133.

Chang, L., and D. Y. Tsao. (2017). The code for facial identity in the primate brain. *Cell, 169*(6), 1013–1028.

Da Costa, A. P., A. E. Leigh, M. S. Man, and K. M. Kendrick. (2004). Face pictures reduce behavioural, autonomic, endocrine and neural indices of stress and fear in sheep. *Proceedings of the Royal Society of London. Series B: Biological Sciences, 271*(1552), 2077–2084.

Knolle, F., R. P. Goncalves, and A. J. Morton. (2017). Sheep recognize familiar and unfamiliar human faces from two-dimensional images. *Royal Society Open Science, 4*(11), 171228.

Kilgour, R. (1981). Use of the Hebb-Williams closed-field test to study the learning ability of Jersey cows. *Animal Behaviour, 29*(3), 850–860.

Veissier, I., A. R. De La Fe, and P. Pradel. (1998). Nonnutritive oral activities and stress responses of veal calves in relation to feeding and housing conditions. *Applied Animal Behaviour Science, 57*(1–2), 35–49.

De la Torre, M. P., E. F. Briefer, B. M. Ochocki, A. G. McElligott, and T. Reader. (2016). Mother–offspring recognition via contact calls in cattle, *Bos taurus*. *Animal Behaviour, 114*, 147–154.

Šárová, R., M. Špinka, I. Stěhulová, F. Ceacero, M. Šimečková, and R. Kotrba. (2013). Pay respect to the elders: age, more than body mass, determines dominance in female beef cattle. *Animal Behaviour, 86*(6), 1315–1323.

Stephenson, M. B., D. W. Bailey, and D. Jensen. (2016). Association patterns of visually-observed cattle on Montana, USA, foothill rangelands. *Applied Animal Behaviour Science, 178*, 7–15.

Howery, L. D., F. D. Provenza, R. E. Banner, and C. B. Scott. (1998). Social and environmental factors influence cattle distribution on rangeland. *Applied Animal Behaviour Science, 55*(3–4), 231–244.

MacKay, J. R., M. J. Haskell, J. M. Deag, and K. van Reenen. (2014). Fear responses to novelty in testing environments are related to day-to-day activity in the home environment in dairy cattle. *Applied Animal Behaviour Science, 152*, 7–16.

Boissy, A., C. Terlouw, and P. le Neindre. (1998). Presence of cues from stressed conspecifics increases reactivity to aversive events in cattle: evidence for the existence of alarm substances in urine. *Physiology and Behavior, 63*(4), 489–495.

Ishiwata, T., R. J. Kilgour, K. Uetake, Y. Eguchi, and T. Tanaka. (2007). Choice of attractive conditions by beef cattle in a Y-maze just after release from restraint. *Journal of Animal Science, 85*(4), 1080–1085.

Laister, S., B. Stockinger, A. M. Regner, K. Zenger, U. Knierim, and C. Winckler. (2011). Social licking in dairy cattle—effects on heart rate in performers and receivers. *Applied Animal Behaviour Science, 130*(3–4), 81–90.

Waiblinger, S., C. Menke, and D. W. Fölsch. (2003). Influences on the avoidance and approach behaviour of dairy cows towards humans on 35 farms. *Applied Animal Behaviour Science, 84*(1), 23–39.

Anthony, L., and G. Spence. (2009). *The elephant whisperer: my life with the herd in the African wild*, vol. 1. Macmillan.

Plotnik, J. M., D. L. Brubaker, R. Dale, L. N. Tiller, H. S. Mumby, and N. S. Clayton. (2019). Elephants have a nose for quantity. *Proceedings of the National Academy of Sciences, 116*(25), 12566–12571.

Bates, L. A., K. N. Sayialel, N. W. Njiraini, C. J. Moss, J. H. Poole, and R. W. Byrne. (2007). Elephants classify human ethnic groups by odor and garment color. *Current Biology, 17*(22), 1938–1942.

Payne, K. B., W. R. Langbauer, and E. M. Thomas. (1986). Infrasonic calls of the Asian elephant (*Elephas maximus*). *Behavioral Ecology and Sociobiology, 18*(4), 297–301.

McComb, K., D. Reby, L. Baker, C. Moss, and S. Sayialel. (2003). Long-distance communication of acoustic cues to social identity in African elephants. *Animal Behaviour, 65*(2), 317–329.

McComb, K., C. Moss, S. Sayialel, and L. Baker. (2000). Unusually extensive networks of vocal recognition in African elephants. *Animal Behaviour, 59*(6), 1103–1109.

Foley, C., N. Pettorelli, and L. Foley. (2008). Severe drought and calf survival in elephants. *Biology Letters, 4*(5), 541–544.

Fishlock, V., C. Caldwell, and P. C. Lee. (2016). Elephant resource-use traditions. *Animal Cognition, 19*(2), 429–433.

McComb, K., G. Shannon, S. M. Durant, K. Sayialel, R. Slotow, J. Poole, and C. Moss. (2011). Leadership in elephants: the adaptive value of age. *Proceedings of the Royal Society B: Biological Sciences, 278*(1722), 3270–3276.

Lahdenperä, M., K. U. Mar, and V. Lummaa. (2016). Nearby grandmother enhances calf survival and reproduction in Asian elephants. *Scientific Reports, 6*(1), 1–10.

Moss, C. J., H. Croze, and P. C. Lee (eds.). (2011). *The Amboseli elephants: a long-term perspective on a long-lived mammal.* University of Chicago Press.

Rasmussen, L. E. L., and V. Krishnamurthy. (2000). How chemical signals integrate Asian elephant society: the known and the unknown. *Zoo Biology, 19*(5), 405–423.

Chiyo, P. I., E. A. Archie, J. A. Hollister-Smith, P. C. Lee, J. H. Poole, C. J. Moss, and S. C. Alberts. (2011). Association patterns of African elephants in all-male groups: the role of age and genetic relatedness. *Animal Behaviour, 81*(6), 1093–1099.

O'Connell-Rodwell, C. E., J. D. Wood, C. Kinzley, T. C. Rodwell, C. Alarcon, S. K. Wasser, and R. Sapolsky. (2011). Male African elephants (*Loxodonta africana*) queue when the stakes are high. *Ethology Ecology and Evolution, 23*(4), 388–397.

Hart, B. L., L. A. Hart, and N. Pinter-Wollman. (2008). Large brains and cognition: Where do elephants fit in? *Neuroscience and Biobehavioral Reviews, 32*(1), 86–98.

Shoshani, J., and J. F. Eisenberg. (1992). Intelligence and survival. In J. Shoshani and F. Knight (eds.), *Elephants: majestic creatures of the wild,* 134–137. Rodale Press.

CHAPTER 7: BLOOD'S THICKER THAN WATER

Heinsohn, R., and C. Packer. (1995). Complex cooperative strategies in group-territorial African lions. *Science, 269*(5228), 1260–1262.

Riedman, M. L. (1982). The evolution of alloparental care and adoption in mammals and birds. *Quarterly Review of Biology, 57*(4), 405–435.

Rudnai, J. A. (2012). *The social life of the lion: a study of the behaviour of wild lions* (Panthera leo massaica [Newmann]) in the Nairobi National Park, Kenya. Springer Science and Business Media.

Funston, P. J., M. G. L. Mills, and H. C. Biggs. (2001). Factors affecting the hunting success of male and female lions in the Kruger National Park. *Journal of Zoology, 253*(4), 419–431.

Stander, P. E., and S. D. Albon. (1993). Hunting success of lions in a semi-arid environment. *Symposia of the Zoological Society of London, 65,* 127–143.

Stander, P. E. (1992). Cooperative hunting in lions: the role of the individual. *Behavioral Ecology and Sociobiology, 29*(6), 445–454.

Smith, J. E., S. K. Memenis, and K. E. Holekamp. (2007). Rank-related partner choice in the fission–fusion society of the spotted hyena (*Crocuta crocuta*). *Behavioral Ecology and Sociobiology, 61*(5), 753–765.

Smith, J. E., R. C. van Horn, K. S. Powning, A. R. Cole, K. E. Graham, S. K. Memenis, and K. E. Holekamp. (2010). Evolutionary forces favoring intragroup coalitions among spotted hyenas and other animals. *Behavioral Ecology, 21*(2), 284–303.

French, J. A., A. C. Mustoe, J. Cavanaugh, and A. K. Birnie. (2013). The influence of androgenic steroid hormones on female aggression in "atypical" mammals. *Philosophical Transactions of the Royal Society B: Biological Sciences*, *368*(1631), 20130084.

Van Horn, R. C., A. L. Engh, K. T. Scribner, S. M. Funk, and K. E. Holekamp. (2004). Behavioural structuring of relatedness in the spotted hyena (*Crocuta crocuta*) suggests direct fitness benefits of clan-level cooperation. *Molecular Ecology*, *13*(2), 449–458.

Theis, K. R., A. Venkataraman, J. A. Dycus, K. D. Koonter, E. N. Schmitt-Matzen, A. P. Wagner, K. E. Holekamp, and T. M. Schmidt. (2013). Symbiotic bacteria appear to mediate hyena social odors. *Proceedings of the National Academy of Sciences*, *110*(49), 19832–19837.

Burgener, N., M. L. East, H. Hofer, and M. Dehnhard. (2008). Do spotted hyena scent marks code for clan membership? In J. L. Hurst, R. Beynon, S. C. Roberts, and T. Wyatt (eds.), *Chemical signals in vertebrates 11*, 169–177. Springer.

Van Horn, R. C., A. L. Engh, K. T. Scribner, S. M. Funk, and K. E. Holekamp. (2004). Behavioural structuring of relatedness in the spotted hyena (*Crocuta crocuta*) suggests direct fitness benefits of clan-level cooperation. *Molecular Ecology*, *13*(2), 449–458.

Drea, C. M., and A. N. Carter. (2009). Cooperative problem solving in a social carnivore. *Animal Behaviour*, *78*(4), 967–977.

Molnar, B., J. Fattebert, R. Palme, P. Ciucci, B. Betschart, D. W. Smith, and P. A. Diehl. (2015). Environmental and intrinsic correlates of stress in free-ranging wolves. *PLOS One*, *10*(9), e0137378.

Coppinger, R., and L. Coppinger. (2001). *Dogs: a startling new understanding of canine origin, behavior and evolution*. Simon and Schuster.

Pierotti, R. J., and B. R. Fogg. (2017). *The first domestication: how wolves and humans coevolved*. Yale University Press.

Hare, B., and M. Tomasello. (2005). Human-like social skills in dogs? *Trends in Cognitive Sciences*, *9*(9), 439–444.

Hare, B., I. Plyusnina, N. Ignacio, O. Schepina, A. Stepika, R. Wrangham, and L. Trut. (2005). Social cognitive evolution in captive foxes is a correlated by-product of experimental domestication. *Current Biology*, *15*(3), 226–230.

Hare, B., and V. Woods. (2013). *The genius of dogs: discovering the unique intelligence of man's best friend*. Simon and Schuster.

CHAPTER 8: CODAS AND CULTURES

Lockyer, C. (1981). Growth and energy budgets of large baleen whales from the Southern Hemisphere. *Food and Agriculture Organization*, *3*, 379–487.

Whitehead, H. (2018). Sperm whale: *Physeter macrocephalus*. In B. Würsig, J. G. M. Thewissen, and K. M. Kovacs (eds.), *Encyclopedia of marine mammals*, 3rd ed., 919–925. Academic Press.

Benoit-Bird, K. J., W. W. Au, and R. Kastelein. (2006). Testing the odontocete acoustic prey debilitation hypothesis: no stunning results. *Journal of the Acoustical Society of America*, *120*(2), 1118–1123.

Fais, A., M. Johnson, M. Wilson, N. A. Soto, and P. T. Madsen. (2016). Sperm whale predator-prey interactions involve chasing and buzzing, but no acoustic stunning. *Scientific Reports, 6*(1), 1–13.

Watkins, W. A., and W. E. Schevill. (1977). Sperm whale codas. *Journal of the Acoustical Society of America, 62*(6), 1485–1490.

Gero, S., H. Whitehead, and L. Rendell. (2016). Individual, unit and vocal clan level identity cues in sperm whale codas. *Royal Society Open Science, 3*(1), 150372.

Konrad, C. M., T. R. Frasier, H. Whitehead, and S. Gero. (2019). Kin selection and allocare in sperm whales. *Behavioral Ecology, 30*(1), 194–201.

Ortega-Ortiz, J. G., D. Engelhaupt, M. Winsor, B. R. Mate, and A. Rus Hoelzel. (2012). Kinship of long-term associates in the highly social sperm whale. *Molecular Ecology, 21*(3), 732–744.

Pitman, R. L., L. T. Ballance, S. I. Mesnick, and S. J. Chivers. (2001). Killer whale predation on sperm whales: observations and implications. *Marine Mammal Science, 17*(3), 494–507.

Curé, C., R. Antunes, A. C. Alves, F. Visser, P. H. Kvadsheim, and P. J. Miller. (2013). Responses of male sperm whales (*Physeter macrocephalus*) to killer whale sounds: implications for anti-predator strategies. *Scientific Reports, 3*(1), 1–7.

Durban, J. W., H. Fearnbach, D. G. Burrows, G. M. Ylitalo, and R. L. Pitman. (2017). Morphological and ecological evidence for two sympatric forms of Type B killer whale around the Antarctic Peninsula. *Polar Biology, 40*(1), 231–236.

Visser, I. N. (1999). A summary of interactions between orca (*Orcinus orca*) and other cetaceans in New Zealand waters. *New Zealand Natural Science 24*, 101–112.

Pyle, P., M. J. Schramm, C. Keiper, and S. D. Anderson. (1999). Predation on a white shark (*Carcharodon carcharias*) by a killer whale (*Orcinus orca*) and a possible case of competitive displacement. *Marine Mammal Science, 15*(2), 563–568.

Baird, R. W., and L. M. Dill. (1996). Ecological and social determinants of group size in transient killer whales. *Behavioral Ecology, 7*(4), 408–416.

Foster, E. A., D. W. Franks, S. Mazzi, S. K. Darden, K. C. Balcomb, J. K. Ford, and D. P. Croft. (2012). Adaptive prolonged postreproductive life span in killer whales. *Science, 337*(6100), 1313.

Wright, B. M., E. H. Stredulinsky, G. M. Ellis, and J. K. Ford. (2016). Kin-directed food sharing promotes lifetime natal philopatry of both sexes in a population of fish-eating killer whales, *Orcinus orca. Animal Behaviour, 115*, 81–95.

Connor, R. C., M. R. Heithaus, and L. M. Barre. (2001). Complex social structure, alliance stability and mating access in a bottlenose dolphin "super-alliance." *Proceedings of the Royal Society of London. Series B: Biological Sciences, 268*(1464), 263–267.

Sakai, M., T. Morisaka, K. Kogi, T. Hishii, and S. Kohshima. (2010). Fine-scale analysis of synchronous breathing in wild Indo-Pacific bottlenose dolphins (*Tursiops aduncus*). *Behavioural Processes, 83*(1), 48–53.

Fellner, W., G. B. Bauer, S. A. Stamper, B. A. Losch, and A. Dahood. (2013). The development of synchronous movement by bottlenose dolphins (*Tursiops truncatus*). *Marine Mammal Science, 29*(3), E203–E225.

Tamaki, N., T. Morisaka, and M. Taki. (2006). Does body contact contribute towards repairing relationships? The association between flipper-rubbing and aggressive behavior in captive bottlenose dolphins. *Behavioural Processes*, *73*(2), 209–215.

Fripp, D., C. Owen, E. Quintana-Rizzo, A. Shapiro, K. Buckstaff, K. Jankowski, R. Wells, and P. Tyack. (2005). Bottlenose dolphin (*Tursiops truncatus*) calves appear to model their signature whistles on the signature whistles of community members. *Animal Cognition*, *8*(1), 17–26.

King, S. L., H. E. Harley, and V. M. Janik. (2014). The role of signature whistle matching in bottlenose dolphins, *Tursiops truncatus*. *Animal Behaviour*, *96*, 79–86.

King, S. L., and V. M. Janik. (2013). Bottlenose dolphins can use learned vocal labels to address each other. *Proceedings of the National Academy of Sciences*, *110*(32), 13216–13221.

Janik, V. M., and P. J. Slater. (1998). Context-specific use suggests that bottlenose dolphin signature whistles are cohesion calls. *Animal Behaviour*, *56*(4), 829–838.

Blomqvist, C., I. Mello, and M. Amundin. (2005). An acoustic play-fight signal in bottlenose dolphins (*Tursiops truncatus*) in human care. *Aquatic Mammals*, *31*(2), 187–194.

Blomqvist, C., and M. Amundin. (2004). High-frequency burst-pulse sounds in agonistic/aggressive interactions in bottlenose dolphins, *Tursiops truncatus*. In J. A. Thomas, C. F. Moss, and M. Vater (eds.), *Echolocation in Bats and Dolphins*, 425–431. University of Chicago Press.

King, S. L., and V. M. Janik. (2015). Come dine with me: food-associated social signalling in wild bottlenose dolphins (*Tursiops truncatus*). *Animal Cognition*, *18*(4), 969–974.

Ridgway, S. H., P. W. Moore, D. A. Carder, and T. A. Romano. (2014). Forward shift of feeding buzz components of dolphins and belugas during associative learning reveals a likely connection to reward expectation, pleasure and brain dopamine activation. *Journal of Experimental Biology*, *217*(16), 2910–2919.

McCowan, B., and D. Reiss. (1995). Whistle contour development in captive-born infant bottlenose dolphins (*Tursiops truncatus*): role of learning. *Journal of Comparative Psychology*, *109*(3), 242.

Schultz, K. W., D. H. Cato, P. J. Corkeron, and M. M. Bryden. (1995). Low frequency narrow-band sounds produced by bottlenose dolphins. *Marine Mammal Science*, *11*(4), 503–509.

Herzing, D. L. (1996). Vocalizations and associated underwater behavior of free-ranging Atlantic spotted dolphins, *Stenella frontalis*, and bottlenose dolphins, *Tursiops truncatus*. *Aquatic Mammals*, *22*, 61–80.

Dos Santos, M. E., S. Louro, M. Couchinho, and C. Brito. (2005). Whistles of bottlenose dolphins (*Tursiops truncatus*) in the Sado Estuary, Portugal: characteristics, production rates, and long-term contour stability. *Aquatic Mammals*, *31*(4), 453.

Kassewitz, J., M. T. Hyson, J. S. Reid, and R. L. Barrera. (2016). A phenomenon discovered while imaging dolphin echolocation sounds. *Journal of Marine Science: Research and Development*, *6*(202), 2.

Sargeant, B. L., and J. Mann. (2009). Developmental evidence for foraging traditions in wild bottlenose dolphins. *Animal Behaviour*, *78*(3), 715–721.

Mann, J., M. A. Stanton, E. M. Patterson, E. J. Bienenstock, and L. O. Singh. (2012). Social networks reveal cultural behaviour in tool-using dolphins. *Nature Communications*, *3*(1), 1–8.

Bender, C. E., D. L. Herzing, and D. F. Bjorklund. (2009). Evidence of teaching in Atlantic spotted dolphins (*Stenella frontalis*) by mother dolphins foraging in the presence of their calves. *Animal Cognition*, *12*(1), 43–53.

Whitehead, H. (2009). Culture in whales and dolphins. In W. F. Perrin, B. Würsig, and J. G. M. Thewissen (eds.), *Encyclopedia of marine mammals*, 2nd ed., 292–294. Academic Press.

Allen, J. A., E. C. Garland, R. A. Dunlop, and M. J. Noad. (2018). Cultural revolutions reduce complexity in the songs of humpback whales. *Proceedings of the Royal Society B*, *285*(1891), 20182088.

Hain, J. H., G. R. Carter, S. D. Kraus, C. A. Mayo, and H. E. Winn. (1982). Feeding behavior of the humpback whale, *Megaptera novaeangliae*, in the western North Atlantic. *Fishery Bulletin*, *80*(2), 259–268.

Allen, J., M. Weinrich, W. Hoppitt, and L. Rendell. (2013). Network-based diffusion analysis reveals cultural transmission of lobtail feeding in humpback whales. *Science*, *340*(6131), 485–488.

Capella, J. J., F. Félix, L. Flórez-González, J. Gibbons, B. Haase, and H. M. Guzman. (2018). Geographic and temporal patterns of non-lethal attacks on humpback whales by killer whales in the eastern South Pacific and the Antarctic Peninsula. *Endangered Species Research*, *37*, 207–218.

Mehta, A. V., J. M. Allen, R. Constantine, C. Garrigue, B. Jann, C. Jenner, M. Marx, et al. (2007). Baleen whales are not important as prey for killer whales *Orcinus orca* in high-latitude regions. *Marine Ecology Progress Series*, *348*, 297–307.

Pitman, R. L., J. A. Totterdell, H. Fearnbach, L. T. Ballance, J. W. Durban, and H. Kemps. (2015). Whale killers: prevalence and ecological implications of killer whale predation on humpback whale calves off Western Australia. *Marine Mammal Science*, *31*(2), 629–657.

Chittleborough, R. G. (1953). Aerial observations on the humpback whale, *Megaptera nodosa* (Bonnaterre), with notes on other species. *Marine and Freshwater Research*, *4*(2), 219–226.

Pitman, R. L., V. B. Deecke, C. M. Gabriele, M. Srinivasan, N. Black, J. Denkinger, J. W. Durban, et al. (2017). Humpback whales interfering when mammal-eating killer whales attack other species: mobbing behavior and interspecific altruism? *Marine Mammal Science*, *33*(1), 7–58.

CHAPTER 9: WAR AND PEACE

Palmour, R. M., J. Mulligan, J. J. Howbert, and F. Ervin. (1997). Of monkeys and men: vervets and the genetics of human-like behaviors. *American Journal of Human Genetics*, *61*(3), 481–488.

Cheney, D. L., and R. M. Seyfarth. (1985). Vervet monkey alarm calls: manipulation through shared information? *Behaviour*, *94*(1–2), 150–166.

Filippi, P., J. V. Congdon, J. Hoang, D. L. Bowling, S. A. Reber, A. Pašukonis, M. Hoeschele, et al. (2017). Humans recognize emotional arousal in vocalizations across all classes of terrestrial vertebrates: evidence for acoustic universals. *Proceedings of the Royal Society B: Biological Sciences*, *284*(1859), 20170990.

Gil-da-Costa, R., A. Braun, M. Lopes, M. D. Hauser, R. E. Carson, P. Herscovitch, and A. Martin. (2004). Toward an evolutionary perspective on conceptual representation: species-specific calls activate visual and affective processing systems in the macaque. *Proceedings of the National Academy of Sciences*, *101*(50), 17516–17521.

Burns-Cusato, M., B. Cusato, and A. C. Glueck. (2013). Barbados green monkeys (*Chlorocebus sabaeus*) recognize ancestral alarm calls after 350 years of isolation. *Behavioural Processes*, *100*, 197–199.

Cheney, D. L., and R. M. Seyfarth. (1988). Assessment of meaning and the detection of unreliable signals by vervet monkeys. *Animal Behaviour*, *36*(2), 477–486.

Byrne, R. W., and A. Whiten. (1985). Tactical deception of familiar individuals in baboons (*Papio ursinus*). *Animal Behaviour 33*(2), 669–673.

Bercovitch, F. B. (1995). Female cooperation, consortship maintenance, and male mating success in savanna baboons. *Animal Behaviour*, *50*(1), 137–149.

Engh, A. L., J. C. Beehner, T. J. Bergman, P. L. Whitten, R. R. Hoffmeier, R. M. Seyfarth, and D. L. Cheney. (2006). Female hierarchy instability, male immigration and infanticide increase glucocorticoid levels in female chacma baboons. *Animal Behaviour*, *71*(5), 1227–1237.

Silk, J. B., J. Altmann, and S. C. Alberts. (2006). Social relationships among adult female baboons (*Papio cynocephalus*). I. Variation in the strength of social bonds. *Behavioral Ecology and Sociobiology*, *61*(2), 183–195.

Archie, E. A., J. Tung, M. Clark, J. Altmann, and S. C. Alberts. (2014). Social affiliation matters: both same-sex and opposite-sex relationships predict survival in wild female baboons. *Proceedings of the Royal Society B: Biological Sciences*, *281*(1793), 20141261.

Städele, V., E. R. Roberts, B. J. Barrett, S. C. Strum, L. Vigilant, and J. B. Silk. (2019). Male–female relationships in olive baboons (*Papio anubis*): Parenting or mating effort? *Journal of Human Evolution*, *127*, 81–92.

Nguyen, N., R. C. van Horn, S. C. Alberts, and J. Altmann. (2009). "Friendships" between new mothers and adult males: adaptive benefits and determinants in wild baboons (*Papio cynocephalus*). *Behavioral Ecology and Sociobiology*, *63*(9), 1331–1344.

Huchard, E., A. Alvergne, D. Féjan, L. A. Knapp, G. Cowlishaw, and M. Raymond. (2010). More than friends? Behavioural and genetic aspects of heterosexual associations in wild chacma baboons. *Behavioral Ecology and Sociobiology*, *64*(5), 769–781.

Baniel, A., G. Cowlishaw, and E. Huchard. (2018). Jealous females? Female competition and reproductive suppression in a wild promiscuous primate. *Proceedings of the Royal Society B: Biological Sciences*, *285*(1886), 20181332.

Silk, J. B., J. C. Beehner, T. J. Bergman, C. Crockford, A. L. Engh, L. R. Moscovice, R. M. Wittig, R. M. Seyfarth, and D. L. Cheney. (2010). Female chacma baboons form strong, equitable, and enduring social bonds. *Behavioral Ecology and Sociobiology*, *64*(11), 1733–1747.

Silk, J. B., D. Rendall, D. L. Cheney, and R. M. Seyfarth. (2003). Natal attraction in adult female baboons (*Papio cynocephalus ursinus*) in the Moremi Reserve, Botswana. *Ethology*, *109*(8), 627–644.

Dart, R. A. (1965). Ahla, the female baboon goatherd. *South African Journal of Science*, *61*(9), 319–324.

Wittig, R. M., C. Crockford, E. Wikberg, R. M. Seyfarth, and D. L. Cheney. (2007). Kin-mediated reconciliation substitutes for direct reconciliation in female baboons. *Proceedings of the Royal Society B: Biological Sciences*, *274*(1613), 1109–1115.

Cheney, D. L., and R. M. Seyfarth. (1999). Recognition of other individuals' social relationships by female baboons. *Animal Behaviour*, *58*(1), 67–75.

Goodall, J. (2010). *Through a window: my thirty years with the chimpanzees of Gombe.* HMH.

Wilson, M. L., C. Boesch, B. Fruth, T. Furuichi, I. C. Gilby, C. Hashimoto, C. L. Hobaiter, et al. (2014). Lethal aggression in Pan is better explained by adaptive strategies than human impacts. *Nature*, *513*(7518), 414–417.

Ladygina-Kots, N. N., F. B. de Waal, and B. Vekker. (2002). *Infant chimpanzee and human child: a classic 1935 comparative study of ape emotions and intelligence.* Oxford University Press.

Crockford, C., R. M. Wittig, K. Langergraber, T. E. Ziegler, K. Zuberbühler, and T. Deschner. (2013). Urinary oxytocin and social bonding in related and unrelated wild chimpanzees. *Proceedings of the Royal Society B: Biological Sciences*, *280*(1755), 20122765.

Whiten, A., and K. Arnold. (2003). Grooming interactions among the chimpanzees of the Budongo Forest, Uganda: tests of five explanatory models. *Behaviour*, *140*(4), 519–552.

Pruetz, J. D., P. Bertolani, K. B. Ontl, S. Lindshield, M. Shelley, and E. G. Wessling. (2015). New evidence on the tool-assisted hunting exhibited by chimpanzees (*Pan troglodytes verus*) in a savannah habitat at Fongoli, Sénégal. *Royal Society Open Science*, *2*(4), 140507.

O'Malley, R. C., W. Wallauer, C. M. Murray, and J. Goodall. (2012). The appearance and spread of ant fishing among the Kasekela chimpanzees of Gombe: a possible case of intercommunity cultural transmission. *Current Anthropology*, *53*(5), 650–663.

Foster, M. W., I. C. Gilby, C. M. Murray, A. Johnson, E. E. Wroblewski, and A. E. Pusey. (2009). Alpha male chimpanzee grooming patterns: implications for dominance "style." *American Journal of Primatology*, *71*(2), 136–144.

Muller, M. N., and R. W. Wrangham. (2004). Dominance, cortisol and stress in wild chimpanzees (*Pan troglodytes schweinfurthii*). *Behavioral Ecology and Sociobiology*, *55*(4), 332–340.

Pruetz, J. D., K. B. Ontl, E. Cleaveland, S. Lindshield, J. Marshack, and E. G. Wessling. (2017). Intragroup lethal aggression in West African chimpanzees (*Pan troglodytes verus*): inferred killing of a former alpha male at Fongoli, Senegal. *International Journal of Primatology*, *38*(1), 31–57.

Lehmann, J., and C. Boesch. (2008). Sexual differences in chimpanzee sociality. *International Journal of Primatology*, *29*(1), 65–81.

Proctor, D. P., S. P. Lambeth, S. J. Schapiro, and S. F. Brosnan. (2011). Male chimpanzees' grooming rates vary by female age, parity, and fertility status. *American Journal of Primatology*, *73*(10), 989–996.

Townsend, S. W., T. Deschner, and K. Zuberbühler. (2008). Female chimpanzees use copulation calls flexibly to prevent social competition. *PLOS One, 3*(6), e2431.

Hopper, L. M., S. J. Schapiro, S. P. Lambeth, and S. F. Brosnan. (2011). Chimpanzees' socially maintained food preferences indicate both conservatism and conformity. *Animal Behaviour, 81*(6), 1195–1202.

Suchak, M., T. M. Eppley, M. W. Campbell, R. A. Feldman, L. F. Quarles, and F. B. de Waal. (2016). How chimpanzees cooperate in a competitive world. *Proceedings of the National Academy of Sciences, 113*(36), 10215–10220.

Furuichi, T. (2011). Female contributions to the peaceful nature of bonobo society. *Evolutionary Anthropology: Issues, News, and Reviews, 20*(4), 131–142.

Surbeck, M., R. Mundry, and G. Hohmann. (2011). Mothers matter! Maternal support, dominance status and mating success in male bonobos (*Pan paniscus*). *Proceedings of the Royal Society B: Biological Sciences, 278*(1705), 590–598.

Surbeck, M., and G. Hohmann. (2017). Affiliations, aggressions and an adoption: male–male relationships in wild bonobos. In B. Hare and S. Yamamoto (eds.), *Bonobos: unique in mind, brain and behaviour,* 35–46. Oxford University Press.

INDEX

Ashley Ward is professor and director of the Animal Behavior Lab at the University of Sydney, where he researches social behavior, learning, and communication across the animal kingdom. His research has been published in leading journals, including *Proceedings of the National Academy of Sciences*, *Biological Reviews*, and *Current Biology*. He lives in Sydney.